URBAN DESIGN FOR WATERFRONT
ADAPTATIONS TO EXTREME SEA LEVEL EVENTS

Wang Liangliang

图书在版编目（CIP）数据

应对极端海平面灾害的滨海区城市设计 = Urban Design for Waterfront Adaptations to Extreme Sea Level Events：英文 / 王量量著 . —北京：中国建筑工业出版社，2017.7

ISBN 978-7-112-21088-6

Ⅰ.①应… Ⅱ.①王… Ⅲ.①滨海－城市规划－建筑设计－研究－英文 Ⅳ.①TU984.183

中国版本图书馆CIP数据核字（2017）第184449号

责任编辑：率　琦　姚丹宁
责任校对：焦　乐　李美娜

Urban Design for Waterfront Adaptations to Extreme Sea Level Events

应对极端海平面灾害的滨海区城市设计

王量量　著
Wang Liangliang

*

中国建筑工业出版社出版、发行（北京海淀三里河路9号）
各地新华书店、建筑书店经销
北京京点图文设计有限公司制版
北京建筑工业印刷厂印刷

*

开本：787×1092毫米　1/16　印张：12¾　字数：303千字
2017年7月第一版　2017年7月第一次印刷
定价：48.00元
ISBN 978-7-112-21088-6
（30740）

版权所有　翻印必究
如有印装质量问题，可寄本社退换
（邮政编码 100037）

Preface and Acknowledgement

Due to the rising sea level and changing climate, extreme sea level events (ESLEs),such as storm surge floods and tsunamis may largely increase in terms of frequency and intensity;and waterfront developments might be seriously threatened by coastal disasters. With the perspective combined with waterfront development, urban sustainability, and climate change, this book examines the effectiveness of the urban form-based adaptations to extreme sea level events proposed by previous studies.

Computational fluid dynamic (CFD) simulations are employed in this research as the main method to examine the hypotheses. Four urban form-based adaptations are theoretically verified to be effective to reduce the impact on buildings during ESLEs. In addition, in order to further test their effectiveness and feasibility an application case study is designed based on the waterfront development proposal for Marina South, Singapore.

One potential contribution of this book is that the research methodology can be adopted by other studies related to of urban form-based adaptation to ESLEs. This research expands the study of waterfront development in aspects of sustainable urban design. With the intention to enhance waterfront resiliency in calamitous climate, this study also contributes some knowledgeto the physical planning and urban design in waterfronts in regard to physical response to ESLEs. In addition, this research examines Singapore's response strategies to sea level rise and related extreme events through the case study of Marina South. The research could provide potential design and assessment strategies to local urban planning authorities.

Being an exploratory study, this research has some limitations. Firstly, research is to compare the resiliency of different urban forms under a same hypothetical condition of ESLEs, the calibration test is omitted. Secondly, due to the limitation of the software and hardware, the simulation durations of some simulation with large scale are comparatively short. Thereby, more comprehensive and multi-variables comparisons and analyses will be developed in further study. In addition, several factors were not considered, such as the effect of drainage system and vegetation in simulations of case studies. Besides, the buildings in all models are simplified as impervious cubes. Therefore, there is a gap for future study to reveal the different capacities of various building shapes, scales, and permeability in response to ES-LEs.

It would have been impossible for me to complete this book without the support and assistance of a number of people to whom I own a debt of gratitude. First and foremost, I would like to acknowledge a deep debt of gratitude to my supervisor, Professor Heng Chye Kiang, who continuously inspired me and supported me to carry out this research. It is a privilege to work with such a meticulous scholar, and I am grateful for his guidance, patience and encouragement during the entire study period. I would also like to thank my thesis committee members, Professor Cheong Hin Fatt and Dr. Nirmal Kishnani, whose constructive suggestions helped shape my initial research, and broadened my academic knowledge. I would like to express my appreciation to Dr. Zhang Ji, Dr. Chen Yu, and A/P Zhu Jieming for their kind help and support, especially the invaluable comments and advice at several critical moments of my research.

I would like to thank the people who help me in CFD simulations, especially Mr. David Welsh, Mr. Nobuyuki Oshitani and Mr. Yuya Ando in the Support Team of the Cradle North America, Madam Man Mei Ling, Mr. Sek Siak Chiang, and Ms. Tan Puay Yong in the Department of Architecture, and Dr. Wang Junhong in Computer Center. Without your technical supports, the simulations in my research would not be realized. Special thanks to Daniel HII Jun Chung and Zhang Xiaofeng, who helped me to carry out a large number of simulations with your knowledge regarding CFD and coastal engineering.

I would like to thank the National University of Singapore, School of Design and Environment, and the Department of Architecture for providing me with scholarship and all kinds of academic supports to facilitate my research.

Special thanks go to the senior members of CASA, especially Chamari Edirisinghe, Doina Andreea Ilies, and Yeo Su-Jan who assisted me with my thesis writing and gave me a great many useful comments about my research. I also would like to thank Katherine Chong Kwang Ping, who provided instructions and assistance to me during my PhD study. Yours kindness and help are sincerely appreciated.

I would also like to acknowledge my spiritual friends, Ms. Chew Moh Leen who is also my English tutor, Dr. Chuang Shaw Choon, Dr. Lou Liang, Dr. Zhong Xin, Li Baoxin, Li Xiaoxi, Zhang Jingyuan and Liu Xiaoli. Your encouragements and prayers are the impetus to my research.

Last but not the least, I owe my deepest gratitude to my parents and my sister for their love and supports which encourage me throughout my study. Particularly, I will dedicate this dissertation to my beautiful and kind-hearted wife, Han Jie, who is always with me in times of difficulties and depressions. My deepest appreciation to you for your deep love, sacrifice and understanding.

Wang Liangliang
National University of Singapore

Contents

Preface and Acknowledgement ·· III

Chapter 1　Introduction ·· 001

Chapter 2　Waterfront developments ·· 009

Chapter 3　Adaptations to extreme sea level events ··············· 030

Chapter 4　Research Methodology ··· 052

Chapter 5　Results and Hypotheses Examination ···················· 072

Chapter 6　Case study and of Singapore ································· 107

Chapter 7　Conclusions ·· 138

References ·· 144

Appendixes ·· 158

Chapter 1   Introduction

Throughout the history of urban development, cities were established adjacent to seas or estuaries in order to acquire accessibility to navigable waters. Besides, protective harbors were also favorable sites for early developments and prosperity (Wrenn, Casazza, & Smart, 1983). The waterfront has always been a unique part in cities for most of coastal cities; however, it is highly threatened by changing climate and rising sea level. With the intention to discover how waterfront development might respond to extreme sea level events, this thesis will discuss the adaptations in the discipline of urban design and planning through simulation experiments and an application study in the city of Singapore. This chapter begins with a brief introduction of the research of waterfront development, which is followed by the discussion of knowledge gap, research objectives, significance, and scope of this research. The structure of the thesis and the overview of the following chapters will also be addressed in the end of this chapter.

1.1   Research background

In the assessment report of Intergovernmental Panel on Climate Change (IPCC), climate change is addressed as an increasingly noticeable phenomenon and evidence such as the observations of increase of global average air and ocean temperatures as well as the widespread melting of glaciers and icecaps have shown that the warming of climate system is unequivocal (Stocker et al., 2013). Currently a common agreement on the causes of climate change has not been achieved. On the one hand, some scholars believe that the observed increase on global average temperatures since the mid-twentieth century is most likely due to the raise in anthropogenic greenhouse gas concentration. They think that our planet has entered the most dangerous era and all the countries should make every effort to reduce the greenhouse gas emission. On the other hand, other researchers believe that global warming is only the natural cycle of the Earth. Scientists with this point of view argue that our earth has gone through several cycles of alternation of warm and cold throughout its existence(Yong, Lee, & Karunaratne, 1991).

Although the main reason for global warming and climate change still remains controversial, the consequences of climate change are quite noticeable and severe as billions of people have already been affected. Among all catastrophic consequences of climate change, the increasing sea level and related extreme events are perhaps the most serious hazard, which are threatening coastal cities. For instance, the 2004 Indian Ocean tsunami killed over 230,000 people in fourteen countries, and inundated coastal communities with waves up to 30 meters high (Paris, Lavigne, Wassmer, & Sartohadi, 2007). And in March 2011, powerful tsunami waves triggered by a 9 magnitude earthquake reached heights of up to 40.5 meters on coastal areas of Japan, which caused tremendous fatalities, injuries, and damages, as well as enormous economic loss (Yalciner et al., 2011).

In a warmer future, climate human settlements could be affected more significantly, as nearly 20 per cent of the world's population lives in areas within 30 km from the sea, and approximately 40 percent live within 100 km from the coast (J. Cohen et al., 1997). A study has estimated that by 2100, 600 million people will inhabit the coastal floodplain below the 1000-year flood level (R. Nicholls & Mimura, 1998). Moreover, the impacts on global economy also cannot be underestimated since coastal cities currently play significant roles in global economy.

Therefore, it is an urgent task to understand how the rising sea and changing climate affect coastal cities' waterfronts. More importantly, planners must upgrade the methods to design and construct waterfronts in vulnerable coastal areas in order to make them more resilient. It is time for the new agendas for waterfront developments to be formulated based on the potential threats. Considering the importance of waterfronts to an island country, this research chose Marina South in Singapore as a case study to discuss the effectiveness and feasibility of the urban design method with adaptations.

1.2 Research context

Ann Breen et al (1994) define waterfront as "the water's edge in cities and towns of all size". According to this definition, waterfront might be an interface area between cities (towns) and water bodies, including rivers, lakes, and seas. Generally, for many coastal cities, the waterfront is the most initial and essential part during their establishments. Therefore, the waterfront development is significant for urban regeneration in terms of keeping the cities' memories. In addition, the economic restructuring and the retreating of industrial and maritime activities encouraged governments to develop waterfronts. For urban planners and developers, this offers the ideal opportunity to re-use prime locations (Minca, 1995). Nowadays local governments are dedicated to developing waterfronts so as to rebuild their urban images (Craig-Smith & Fagence, 1995; Marshall, 2001b).

Along with the evolution of the transportation networks and shipping technology, waterfronts may have lost their industrial and transportation functions and their development processes have become unique phenomena in urban study. Nowadays, waterfronts are the most dynamic and vigorous spaces in cities and also became essential sections of the urban culture context. At a particular stage, waterfronts were rediscovered as an antidote to urban decline and suburbanization, especially in North America. In academia, waterfront development has drawn professional interest and become a research topic since the 1960s (Marshall, 2001b). In recent decades, with the rising sea level and changing climate, the waterfront development might become increasingly important for most coastal cities. On the one hand, waterfront areas accommodate important urban functions, such as harbors, residences, and even centralbusiness districts (CBD), and on the other hand, coastal waterfronts might be influenced more by climate change and sea level rise. Accordingly, providing effective and feasible adaptations for waterfront development has become an important task in order to respond to extreme sea level events.

Adaptations aim to change the current urban planning system in order to reduce the damage due to sea level rise. Response Strategies Working Group and Coastal Zone Management Subgroup (1990) summarized three main broad categories, which include retreat, protection and accommodation. In the following chapters, these adaptations strategies will be specifically addressed. Generally, for coastal cities with limited land, retreat is not an option. Nowadays, most research attention is given to the protection strategies as they are the most popular adaptations and quite effective to respond to seawater corrosion. However, recent studies have shown that some protection strategies are not environmental friendly, especially the hard protections, such as concrete sea walls and barrages. Moreover, for some extreme sea level events like storms, surge floods and tsunamis, protections may not be effective enough to prevent seawater intrusions. Once the protection strategies fail, the consequences caused by extreme sea level events will be more serious if there are no other adaptation strategies applied on waterfronts.

As a coastal city in Asian Pacific region, Singapore is affected by sea level rise and climate change. Although Singapore with a location protected by periphery islands is safer than other coastal cities in this region like Manila, Hong Kong, Shanghai, and Tianjin, its waterfront developments are seriously threatened by sea level rise and related extreme events. With other impacts, coastal land loss and increasing flood risk are identified as two major challenges for urban waterfront developments in the national report which addresses climate change (NCCS, 2008). Therefore, this research chooses a waterfront development in Singapore as an application case study.

1.3 Statement of the research topic

As addressed previously, in order to respond to extreme sea level events, more adaptations must be applied on waterfrontsbesides protection strategies. With the intention to enhance the sustainability and resiliency of waterfront development, a few researchers have focused on urban form-based adaptations in the discipline of urban study(D.M. Bush, O.H. Pilkey, & W.J. Neal, 1996; S. Davoudi, J. Crawford, & A. Mehmood, 2009; K. Walsh et al., 2004; Wilson & Piper, 2010), such as controlling development density and transformingthe urban fabric. Their goals are to integrate the urban design and physical planning of waterfront developments with adaptations in order to make them more resilient. However, most of these studies still remain in hypotheses and theoretical discussions. Therefore, for the purpose of urban planning practices, the effectiveness of these recommended adaptations needs to beverified through scientificmethodologies.

Recently, researchers have carried out some study to predict the possible impacts caused by sea level rise and climate change in the Singapore Strait. Although thevulnerability assessments to the potential impacts of climate change are conducted by research institutes under the coordination of National Environment Authority, urban planning authorities of Singapore, such as the Urban Redevelopment Authority and Building Construction Authority, havenot developed a series of plans to address extreme sea level eventslike most of the coastal cities in developing counties.

1.3.1 Research gaps

Based on the discussion above, it is clear that to propose effective and feasible urban form-based adaptations for waterfront developments is quite an urgent task. With the research perspective combined with urban design and sustainable development, the research gaps of this study are identified as follows:

- There is a need to summarize the challenges and impacts of climate change and sea level rise on waterfront developments.

- Although some urban form-based adaptations to extreme sea level events are hypothetically proposed, their effectiveness and feasibility must be verified through scientific methodologies.

- As a coastal city that highly depends on its waterfronts in terms of economic development and urban image, Singapore has not carried out research on the topic of adaptations to extreme sea level events from the perspective of urban study. Meanwhile, urban planning authorities have not integrated waterfront developments with urban form-based adaptations in order to enhance their resiliency.

1.3.2 Research question

The proposed research question of this study is the following:

Can urban form-based adaptations effectively reduce the impact of extreme sea level events on waterfront developments and can these adaptationsintegrated with the urban design and physical planning of waterfronts?

1.3.3 Research objectives

With the intention to fill the research gaps and to answer the research question, the specific objectives of this research include the following tasks:

- to summarize the impacts of climate change and sea level rise on waterfront development as well as adaptations proposed by recent studies

- to examine and verify the effectivenessof urban form-based adaptations to extreme sea level events analyzed by recent studies

- to integrate the identified effective adaptations with the waterfront development proposal in Singapore and recommend useful design methods that may enhance the capacity to respond to extreme sea level events.

1.3.4 Research hypothesis

The main research hypothesis in this thesis is that urban form-based adaptations can effectively reduce the impact of extreme sea level events and enhance the waterfronts' resiliency and they can be feasibly applied on waterfront developments. As part of the thesis, there are also five specific hypotheses regarding urban forms, which will be particularly addressed in the third chapter, and they are related to development density, urban block typologies, block depth, block orientation, and the spatialcharacteristics of U-shaped blocks.

1.4 Significance of the research

Nowadays, coastal cities are increasingly important to the global economic network and their waterfronts are essential parts of the cities in terms of economic activities, urban image, and residents' living quality. Therefore, under the circumstance of changing climate and rising sea level, it is particularly significant to enhance the resiliency and sustainability for waterfronts.

The results of this study may contribute some knowledge for the physical planning and urban design in waterfronts in terms of response to extreme sea level events. The expected conclusion could provide sustainable strategies for coastal cities and might be very influential inshaping cities' waterfronts. For the new development schemes around coastlines or on reclaimed land, this research could provide a useful assessment methodology for urban planning authorities to identify effective urban designs for waterfront developments, which might be useful for decision makings. For the existing urban areas with vulnerable locations, the feasible and effective adaptations verified in this study could be adopted for local adjustments in order to improve their capacity to deal with extreme sea level events.

Meanwhile, an application case study will be carried out based on the waterfront developments in Singapore. Therefore, its current response strategies to sea level rise and related extreme events might be examined. Consequently, the waterfronts' capacities to deal with climate change can be discussed. Hopefully, effective adaptations verified in this study can be integrated with Singapore's waterfront developments so that the research can provide significant design and assessment strategies to local urban planning authorities.

As urban planners and researchers, it is our obligation and primary target to improve cities' capacities of responding to climate changes. More frequent and fierce coastal disasters challenge our waterfront development, but they also provide opportunities for the transformation of the manner in which we design and manage it. It is believed that this research can help designers and planners in this regard.

1.5 Scope of the study

Generally, the response strategy to ESLEs cannot merely be constituted by urban form-based adaptations. It is usually combined with offshore protection engineering, costal protective structures, integrated coastal managements, and so on. Thus, this research only focuses on the physical dimension of urban design on a meso-scale which is defined as urban blocks.

Obviously, due to the variety of urban structures and topographies of urban waterfronts, the typology of urban blocks might contain too many possibilities, which may be impossible to be considered and analyzed. Therefore, this research is carried out under the framework of a grid-pattern urban layout with flat terrain. According to Lynch and Rodwin(1958), the grid-pattern is the most original and effective urban layout. Rose-Redwood (2008)believe that from urban planning, architecture, to modern art and cartography, a range of human activities are influenced by the grid pattern urban structure. Grid streets patternscan create more regular blocks; most of them are rectangular, which is believed to be easier tobe designed and managed, especially for residential areas (Grammenos, Pogharian, & Tasker-Brown, 2002), and more importantly it might be more adaptive for a compact urban environment.

In terms of the damage caused by extreme sea level events, there are two aspects appearing to be most prominent. One is the destruction of ground and underground infrastructures and street furniture in urban open space while another aspect concerns the impact on buildings in waterfront as they directly bear the strikes. Although it is acknowledged that both aspects are important issues, compared to the former, the latter might be more important due to significant economic losses and threats to social security. Therefore, this research evaluates the effectiveness of adaptations by measuring their impact on buildings in waterfronts.

Meanwhile, a computational fluid dynamic simulation is employed in this research as the main research tool which contains complex physical and mathematical models. The process of simulation may involve some mathematical issues, but these are not essential parts of this research thus beyond the scope of this thesis and will be not discussed.

1.6 Overview of chapters

There are a total of seven chapters in this thesis. The first chapter briefs the study, in which the research gaps, question, objectives and hypothesis are addressed after the discussion of research background and context. The significance and scope of this study are also stated in this chapter. The second chapter mainly focuses on the literature of waterfront. Through a discussion regarding the typical evolution process of waterfront, its development motivations and challenges, various research perspectives for the study of waterfront developments are also included in the second chapter. The third chapter introduces the response strategies to climate change and sea level rise and illustratesfive specific hypotheses of this research. The fourth chapter represents the research methodology and the research framework through Figures and narratives. Meanwhile, five experiments are designed and illustrated in this chapter in order to examine the hypothesis. The fifth chapter reports the results of experiments and gives analyses based on the comparison of experiment results so that the hypotheses may be verified. Although the experiments may theoretically prove some hypotheses, the applicability of verified adaptations may need a case study to prove their feasibility in real waterfront developments. Therefore, in the sixth chapter, an application case study will be addressed with the intention of integrating these adaptations within a waterfront development proposal and of discovering their capacities to reduce the impact of extreme sea level events. The last chapter summarizes the thesis with research conclusions, contributions and limitations. Finally, the thesis will be ended by the recommendation for future studies.

Chapter 2 Waterfront developments

In this chapter, the common evolution process of waterfront will be explained, followed by the discussing the motivations and challenges of waterfront development. This chapter will conclude with a discussion of various research perspective related to waterfront.

2.1 Evolution process of waterfront

It is interesting to notice that most of waterfronts are not initially design for landscape or public space but for industrial or maritime purposes. It would be meaningful to start the discussion of waterfront development by explaining its transformation process. B. S. Hoyle, et al (1988) address waterfront development through the changing relationship between cities and ports, which summarize the typical evolution process. They divided the process into five stages (Figure 2.1), namely the primitive port-city, the expanding port-city, the modern industrial port-city, retreat from the waterfront, and redevelopment of the waterfront (B. S. Hoyle et al., 1988). This section will extend discussions based on their theory.

Stage	Symbol ○ city ● port	Period	Characteristics
I Primitive cityport	◐	Ancient-medieval to 19th century	Close spatial and functional association between city and port
II Expanding cityport	◐--●	19th-early 20th century	Rapid commercial and industrial growth forces port to develop beyond city confines. with linear quays and break-bulk industries
III Modern industrial cityport	○--●	mid-20th century	Industrial growth (especially oil refinning) and introduction of containers and ro-ro facilities require separation and increased space
IV Retreat from the waterfront	◐ ●	1960s-80s	Changes in maritime technology induce growth of separate maritime industrial development areas
V Redevelopment of the waterfront	◐ ●	1970s-90s	Large-scale modern port consumes large areas of land- and water-space; urban renewal of original core

Figure 2.1 Stages in the evolution of the port-city interface
Source: B.S. Hoyle et al. (1988) p.7.

2.1.1 The primitive port-city

Since the age of great discovery, most global cities were established on waterfronts in order to acquire direct or indirect accessibility of navigable waters as the movement of products, raw materials and people were primarily dependent on water transportation. By the beginning of the 18th century, the first colonial settlements of North America were established along the Atlantic coast, such as Boston, Charleston, Baltimore, New Amsterdam (renamed as Now York in 1644) and Philadelphia (Wrenn et al., 1983). Although they were just a small town adjacent to the waterfront at the very start, these cities have become the most important metropolitan centers in North America. Millions of immigrants preferred to settle down there because of better opportunities and a prosperous economy (Figure 2.2).

Figure 2.2 The first colonial settlements of North American along the Atlantic coast
Source: Wrenn, D. M. (1983) p.5.

In Europe, the great medieval Italian port-cities, Genoa, Naples, Venice, were the first to develop a commercial system whereby the products of distant lands became the necessities and luxuries of urban life (B. S. Hoyle et al., 1988). On the other side of the world, in Asia, with the needs of export and import, some coastal area with preferably navigable locations got the opportunities and expanded to larger cities from small towns and fishing villages, such as Singapore, Malacca, Hong Kong, Shanghai, Shenzhen, and so on.

In summary, in the stage of primitive port-city stage, the city and the port were dependent upon each other; and in some cases, they combined and penetrated each other so that the difference was not apparent. The co-existence of a primitive port and city involved close spatial association and maximum functional interdependence (B. S. Hoyle et al., 1988).

2.1.2 The expanding port-city

Rapid commercial and industrial development in the 19th century forced many ports to break out of their traditional confines, and the seeds of port-city separation were widely scattered; at the same time, the expansion of port facilities exerted a marked influence upon patterns of urban land use (B. S. Hoyle et al., 1988). The limited land was not sufficient for the bigger maritime facilities, and the bigger ships also needed deeper water area to berth. While some ports extended forwards into the water through reclamation. In this period, some of the expanding port-cities became effective instruments of exploitation and colonization so that the development process was introduced all over the world (B. S. Hoyle et al., 1988).

2.1.3 The modern industrial port-city

Following the industrial revolution and development of water transport, economic value was further boosted in waterfront areas. Industry zones, harbor zones, warehousing zones and rail stations flocked to areas near water and developed in an intensive way to reduce cost, making waterfront areas the most economically dynamic sections in a city. Meanwhile, waterfront played a decreasing role in urban public life as environmental quality in the areas deteriorated sharply (Meyer, 1999a; Wrenn et al., 1983).

2.1.4 Retreat from the waterfront

The retreat from the traditional location of the ports was largely induced by the development of maritime technology(B. Hoyle, 2000). Since the middle of the 20th century, there was an increasing trend in adjustment of the industry structure throughout the world. As the traditional industries, warehousing industry, harbors and railway stations moved outward from urban downtowns, waterfronts areas once occupied by these major facilities either fell apart or collapsed (B. S. Hoyle et al., 1988). As the global industrial structure adjusted, urban waterfront areas underwent deindustrialization (Hamilton & Simard, 1993; Marshall, 2001b), in which most of industries and harbors separated from urban downtowns, while development of the modern aerospace industry, automobile and railway industries weakened the dominant position of water harbors as hubs of urban transport (Bone, Betts, & Greenberg, 1997).

2.1.5 Redevelopment of the waterfront

As a result of the retreat, the function of waterfront gradually changed, with its function of the maritime and industry steadily reducing and function of landscape boosting. Meanwhile, urban layout also changed, where on the one hand, most industrial facilities were displaced to satellite towns or development zones other than urban downtown; on the other hand, people required a higher quality of urban life, and began to re-realize the potential of urban waterfronts(Har, 1997). In decades, the renaissance of urban culture and history and the movement revitalizing urban waterfronts were unfolded in most of postindustrial cities.

2.1.6 Section summary

The discussion above gives a general idea and is useful to understand how the port-related waterside evolves into urban waterfronts. It is not saying that all the urban waterfronts follow the same development path. While more or less, it might be possible to divide the transformation process into several stages according to B. S. Hoyle, et al.'s (1988) theory. For example, table 2.1 shows most of the significant historic periods during the development process of urban waterfront in Toronto, Canada.

2.2 Motivations of waterfront development

In the previous section, it is addressed that inner city regeneration might be an important reason to carry out waterfront development (B. Hoyle, 1995). According to the statistics, up to thousands of waterfront projects have been developed in North America since 1965, more than 90 waterfront projects were being constructed in the USA alone in 1989 and hundreds of such projects are currently being undertaken all around the world (Breen & Rigby, 1996; B. S. Hoyle et al., 1988). In this section, the question of why local governments are dedicating themselves to such projects will be discussed. Although there are significant differences between various regions, four motivations of waterfront developments are commonly mentioned, namely economic restructuring city image remaking, the growth of awareness about urban environment, the growth of urban tourism and leisure, and suburban sprawl and urban decay.

The development of Toronto waterfront   Table 2.1

Primitive port-city	1793	Established as a town, named York.
Expending port-city	1850s	Became an important commercial port town.
Modern industrial port-city	Later in 1850s	First redevelopment of waterfront began because of the new maritime technology, industry and railway.

Modern industrial port-city	1920s	Large-scale redevelopment of waterfront to be the financial and commercial center. A Comprehensive Waterfront Development Plan(1912) located industry on the waterfront which formulated by Toronto Board of Trade and Toronto Harbor Commission.
	1940s	The 1912 plan had been largely achieved. The force of industrialization development was over.
Retreat from the waterfront	1950s	Suburbanization began. Industry declined. Rail and water transportation declined.
Redevelopment of waterfronts	1969	Official plan of urban waterfronts was adopted to promote more commercial development and attract the middle and upper-middle class back to the urban.
	1978	A new plan City of Toronto 1978 was adopted to redevelop Central Waterfront, to make room for the new commercial activity and to diversify the land use

Source: B.S. HOYLE, et al. (1988) and recogniized by author.

2.2.1 Economic restructuring and city image remaking

The first reason that encourages governments to develop the waterfronts might be the transformation from industrial cities to postindustrial cities. According to Hall (1991), the progressive abandonment of the docklands by port activities started in the 1950s in America, followed by a process of deindustrialization in urban waterfronts happened elsewhere over the following decades. Holy et al (1988) argue that the industrial port-based activities gradually retreat from urban waterfront, which provides a spatial and functional vacuum. For local governments, it is a chance to shift the economic focus from secondary industry to the tertiary one. For urban planners and developers, this was the ideal opportunity to re-use prime locations, both for land-based concerns such as housing, restaurants and shopping centers, and for maritime interests, such as water-based recreation facilities (Minca, 1995).

Besides, in order to attract more investment, a great of attention is given to the remaking of cities' image. Nowadays waterfront has become a special place serving as a main section for demonstration of urban image (Craig-Smith & Fagence, 1995; Marshall, 2001b). According to Kevin Lynch's (1992) theory, cities are imaginable through unique elements, which are defined by Lynch as a network of paths, edges, districts, nodes, and landmarks. While, Jutla (2000) expatiates it as a powerful characteristic through which a city is known to people, which provides a visual identity for a city. In other words, the remaking of the image is to show cities' ambition for the new developing vision (Marshall, 2001b).

According to Serageldin(1997), "competition for market shares in the global economy will force major adjustments to the urban fabrics of cities as they rationalize to realize their economic potentials (p. 40)." Nowadays, the competitive advantages depend less on location than before, and more on the availability of an appropriate infrastructure (Marshall 1998). As probably the most visible locations, waterfronts provide remarkable opportunity for the urban regeneration efforts. Being the sites of the former industrial operations, waterfronts were often the most degraded places of the industrial period. Because of these reasons, waterfront is not only crucial to the cities' developmentbut also essentialto the quality of the urban expression (Marshall, 2001b). In addition, the cities' image is not only relying on the physical environment but also on the culture.

Fairly large number of examples demonstrates the importance of waterfront in the process of remaking the city's image. Many cities around the world attempt to transform cities' economy through redevelopment of waterfronts. However, the remaking of the city image cannot be fulfilled through several projects but requires the regeneration of the whole system, including cleaning the polluted water body, preserving the historic urban fabric, providing connectivity, accessibility, and social activities, etc.

2.2.2　The growth of awareness about urban environment

Water has played a central role in the development of human civilization since its earliest days. The development around waterside locations reached a peak during the industrial revolution and the first half of last century and usually proceeded without regard to the quality of water. Waterfronts became progressively more polluted and the aquatic environment was severely degraded (Meyer, 1999a). However, industrial usage of waterways and docks declined during the second half of last century and because of their poor environmental quality many have been simply abandoned and allowed to deteriorate further (White, 1993).

After the 1950s, in the highly developed counties, waterfronts and other urban places tried to meet the needs created by the new lifestyle, which could be identified with the characteristics of having a great deal more leisure time and being divided into ever-smaller communities than previous generations (Marshall, 2001b). The diversity of the space and activities became an important evaluation standard for urban built environment (Marcus & Francis, 1997). The progressive awareness of the quality of the urban environment forces the local governments to carry out a series of water-related projects. Actually, urban waterfront can provide unique landscape resources its form, texture, and special features(Wrenn, 1983). Therefore, along with the improving of water quality and upgrading the environment, waterfronts would be able to be renovated.

2.2.3 The growth of urban tourism and leisure

According to Craig-Smith and Fagence(1995), urban tourism, leisure, and recreation prefer to the venues with historic value and heritage, which waterfronts could provide the prefect opportunities due to the recession of industry and transportation. In addition, the lifestyle changed dramatically in 1940s, which could attribute to the sudden switch from six-day workweek to five-day workweek and from the two-earner family into the one-earner family (Marshall, 2001b). The demand for travel to cities has greatly increased over the last few decades. The waterfronts have the natural and cultural characters attracting people. Lately, local governments and private investors are paying increasing attention to waterfronts because they pose a variety of opportunities for tourism (Ballentine, 2006). For example, since the Chicago World Exposition in 1893, conventions, exhibitions, festivals and events are held by cities and their waterfronts have become the favored site for these events (Hines & Harris, 2008).

In an era of increased leisure, recreational participation, and increased environmental and heritage concern, many of the world's major waterfronts have been redeveloped with conservation, recreation and tourism in mind. The redecoration and tourism can be cited as catalysts for redevelopment (Craig-Smith & Fagence, 1995).

2.2.4 Suburban sprawl and urban decay

Another phenomenon driving the waterfront redevelopment is suburban sprawl and urban center decay. Although most of these cases happened in developed countries, especially in the United States, they are worthy to be discussed due to their significance. Suburban sprawl and urban decay were the twin problems that deindustrialized and decentralized cities have suffered since the 1950s. The postwar years have been marked by growth in population and employment which has centered on the suburbs (Charlesworth & Cochrane, 1994).According to Stanback and Knight (1976), during 1950 to 1974, in United States, population increased by 56.6 million, 70% was accounted for by growth in suburban rigs, only 14% by central cities. The suburban sprawl and urban decay rustled in the incredible wastage of farmland and degradation of security and environment in urban centers (Calthorpe & Fulton, 2001).

Governments and researchers want to attract people back to urban centers, especially the middle-class. Many cities dedicate themselves to revitalizing the downtown through urban renewal plans. With its attraction, waterfront was regarded as the antidote of the urban decay (B. S. Hoyle et al., 1988). With the important location and splendid landscape resources, old harbors and ports became public spaces, central business districts and high-class residential areas in order to attract people back to the cities.

2.3 Challenges for waterfront development: sea level rise andextreme sea level events

Recently, another issue has made waterfront development more problematical in the form of climate change which has become increasingly visible and serious and is threatening urban development. According to Intergovernmental Panel on Climate Change(IPCC, 2007), the global average temperature is increasing at an unprecedented rate.. As mentioned in the first chapter, as the serious consequences of global warming, sea level rise (SRL) and extreme sea level events (ESLEs), including more frequent storm surges, hurricanes and tsunamis, raise challenges for waterfront development, especially for coastal cities. Waterfront resiliency and coastal safety has been called as the most principal concerns. In order to respond, the first step is to understand how the changing climate and sea level rise affect the waterfront development. In this section, few ideas of sea level rise and its impact on waterfront development are discussed.

2.3.1 Sea level rise

Based on the statistics of tide-gauges around the word, sea level is rising with an increasing rate. According to the latest data released by IPCC, since the mid-19th, the rate of SLR has become larger than the mean rate during the previous 2000 years(Stocker et al., 2013). In this report, the rate of global SLR is summarized in table 2.2. As it shown in this table, it is quite evident that the rate is increasing significantly.

The rate of sea level rise   Table 2.2

Period	Mean rate of global sea level rise (mm/per year)	Ranges of the rate (mm/per year)
1901 – 2010	1.7	1.5 to 1.9
1971 – 2010	2.0	1.7 to 2.3
1993 – 2010	3.2	2.8 to 3.6

Source: Stocker et al. (2013).

1) The contributors of global mean sea level rise

Generally, the global mean sea level rise could be attributed quite a lot of sources, such as the thermal expansion of the oceans, the melting of glaciers and ice caps, the change to the major Greenland and Antarctic sheets, and the sum of individual climate contributions to SLR (tsunamis and storm surge). Table 2.3 summarizes several main contributors and their contributions to global mean SLR. Among all sources, contributions from glacier mass loss and ocean thermal expansion are the most significant ones, which totally explain about 75% of the observed global mean SLR since the early 1970s, and which is also believed as the results of anthropogenic intervention (ibid).

Based on such statistics, it might be concluded that sea level will rise in the next few decades even if the concentration of greenhouse gas in the atmosphere could remain stable right now. In spite of lots of uncertainties, the sensitivity of sea to different scenarios before 2050 is quite small due to the slow reaction of deep seawater and glaciers to global temperature changes. In other words, uncertainties in future emissions of greenhouse gases play a smaller part over the next few decades, as the ocean has a substantial thermal inertia and only responds slowly to external force by greenhouse gases. For example, by 2050, it is suggested that the uncertainties due to emissions alone are only a few centimeters of total sea level. The effects of different emission scenarios become larger towards 2100 and beyond (Church, Gregory et al. 2001).

Observed contributions (mm) and likely range (mm) to global mean sea level (GMSL) rise   Table 2.3

Source	1901–1990	1971–2010	1993–2010
Thermal expansion	–	0.8 [0.5 to 1.1]	1.1 [0.8 to 1.4]
Glaciers except in Greenland and Antarctica	0.54 [0.47 to 0.61]	0.62 [0.25 to 0.99]	0.76 [0.39 to 1.13]
Glaciers in Greenland	0.15 [0.10 to 0.19]	0.06 [0.03 to 0.09]	0.10 [0.07 to 0.13]
Greenland ice sheet	–	–	0.33 [0.25 to 0.41]
Antarctic ice sheet	–	–	0.27 [0.16 to 0.38]
Land water storage	–0.11 [–0.16 to –0.06]	0.12 [0.03 to 0.22]	0.38 [0.26 to 0.49]

Source	1901–1990	1971–2010	1993–2010
Total of contributions	–	–	2.8 [2.3 to 3.4]
Observed GMSL rise	1.5 [1.3 to 1.7]	2.0 [1.7 to 2.3]	3 [2.8 to 3.6]

Source: Stocker et al. (2013).

2) The prediction of future sea level rise

Apart from studies regarding the contributors of SLR, another important issueis to project its trend since it is currently impossible to stabilize sea level. The prediction of SLR trend is currently based on Global Climate Models (GCMs). Although these models contain many uncertainties due to limited understanding of climate change, with the continuous improvement of performances, they are currently the best available projection methods.

The projections of future Global Mean SLR done by——IPCC just release its fifth assessment report for climate change report (AR5), which is summarized in table 2.4.These projections is relative to the Global Mean SLR of 1986–2005 and are based on the analysis of Global Climate Model's calculations and available studies, in which ocean thermal expansion, contributions from glaciers, Greenland and Antarctica and terrestrial contributions are included. These projections vary as the greenhouse gas emission scenarios changes. For scenarios are included, namely RCP2.6, RCP4.5, RCP6.0, and RCP8.5, distinguished in ascending order of greenhouse gas emissions. Both projected increments of temperature and sea level are provided in the table. The central values and the 5–95% range are shown in the sea level projections before 2100; from 2200 onward, only the range are given due to limited available data(Stocker et al., 2013).

Projected change in global mean surface air temperature and global mean sea level rise for the mid– and late 21st century relative to the reference period of 1986–2005.   Table 2.4

		2046–2065		2081–2100	
	Scenario	Mean	Likely range	Mean	Likely range
Global Mean Surface Temperature Change (℃)	RCP2.6	1.0	0.4 to 1.6	1.0	0.3 to 1.7
	RCP4.5	1.4	0.9 to 2.0	1.8	1.1 to 2.6
	RCP6.0	1.3	0.8 to 1.8	2.2	1.4 to 3.1
	RCP8.5	2.0	1.4 to 2.6	3.7	2.6 to 4.8
	Scenario	Mean	Likely range	Mean	Likely range

		2046-2065		2081-2100	
	Scenario	Mean	Likely range	Mean	Likely range
Global Mean Sea Level Rise (m)	RCP2.6	0.24	0.17 to 0.32	0.40	0.26 to 0.55
	RCP4.5	0.26	0.19 to 0.33	0.47	0.32 to 0.63
	RCP6.0	0.25	0.18 to 0.32	0.48	0.33 to 0.63
	RCP8.5	0.30	0.22 to 0.38	0.63	0.45 to 0.82

Note: RCPs is the abbreviation for Representative Concentration Pathways, which are four greenhouse gas concentration (not emissions) trajectories adopted by the IPCC for its fifth Assessment Report.
Source: Stocker, et al. (2013)p. 23.

Although the predictions provided by IPCC is most often quoted data by other researches, it has been criticized to be too optimistic. Some researchers have addressed that sea level rise could reach 1 to 2 meters by the end of this century (Rahmstorf, 2007).

2.3.2 Impacts on waterfront development

From the perspective of waterfront developments, coastal land loss and increasing extreme events could be directly caused by sea level rise, which should be addressed in the framework of urban planning system. Thereby, this thesis will focus on such issues and discuss them as impacts of sea level rise and climate change.

1) Coastal land loss

The biggest threat caused by the rising sea level is coastal land loss as most island states are suffering from it. For example, some Pacific island nations, such as Tuvalu has been on the verge of being submerged. Given the seriousness of this issue, a large number of studies have been carried out and attract many people's attention. Dwarakish, et al. (2009)carry out a study which shows that due to SLR around Udupi Coast in Karnataka state, the rate of erosion was 0.6018 km^2/yr during 2000-2006 and in the total 95 km long beach, 46 km long coastline (59%) is under critical erosion. They also calculate that with 1 meter increase of sea level, 42.19 km^2 lands could submerge by flooding; while 10 meter sea level rise will cost 372.08 km^2 land loss. Another study shows that by 2100 in California's coast, 1.4 meter sea level rise will result in a land loss of 41 square miles and currently 14,000 people is living in the area with danger of future erosion(Heberger, Cooley, Herrera, Gleick, & Moore, 2011).

While in developing counties, especially in Asia, the situation is even worse due to scarcity of effective management of coastal ecosystem and excessive extraction of groundwater resource (Ali, 1996; Chen, 1997; Z Huang, Zong, & Zhang, 2004;Mei-e,1993;W.S. Ng & Mendelsohn, 2006) A 72-year tidal record of Hong Kong (Z Huang et al., 2004) concludes that 30 cm water level rise in Pearl River delta is possible by 2030. Regarding this topic, a more comprehensive review is discussed in Chapter 6 based on the case of Singapore.

2) Increasing risk of ESLEs

In addition to coastal land loss, another major effect caused by sea level rise would be the increasing risks of extreme sea level events (ESLEs). According to Church, Hunter et al. (2006)ESLEs refer to "events are those driven by severe weather such as tropical cyclones and mid-latitude storms." It is not difficult to understand that sea level rise can weaken the defensive capacities of coastal areas by increasing the tidal level. With one to two meters sea level rise, some low-lying coasts could become flood prone areas. Moreover, if the high tide events, such as storm surge and tsunami occur, raging floods might break through the coastal protective structures and cause catastrophic disasters. Regarding such issue, quite a number of researches show their academic interests.

Although it is not a consensus yet that some climate systems (characteristics of ENSO, large-scale atmospheric circulation) will change their behaviors along with global warming, the alterations to sea surface temperatures could be one of the important factors which change the frequency and intensity of tropical cyclones(Walsh & Pittock, 1998). Similarly, IPCC(2007) projects an increase of 10-20% in tropical cyclone intensify if the sea surface temperature would increase by 2-4 °C from current threshold temperature. The amplified storm surge height caused by higher sea surface temperature and low pressure associated with tropical storms might enhanced the risk of coastal disasters(IPCC, 2007).

Meanwhile another research (Nurse et al., 2001)addresses that if the mean sea level rises with all other condition being the same, present extreme sea level events will be attained more frequency, which means that the increase in maximum heights will be equal to the increase in mean sea level. Accordingly, even small increase of sea level would have severely negative effects on atolls and low-lying islands (R. J. Nicholls, Hoozemans, & Marchand, 1999). The prospect of extreme sea level events, such as storm surge or higher wave amplitudes should not be underestimated (Nurse et al., 2001). Nicholls et al (2012)projects that, given a 38 cm increase between 1990 and the 2080s, "many coastal areas are likely to experience annual or more frequent flooding". The pacific oceans will especially face the largest relative increase in flood risk(Field et al., 2012).

In addition, Walsh, Betts et al (2004) address that rising seas may increase the incidence of coastal flooding, either by increasing the height of storm surges, or by acting as a higher seaward barrier which restricts the escape of flood waters caused by excessive runoff.

2.3.3 Section summary

Obviously, with its significance to urban environment and importance in responding climate change and SLR, waterfront development is attracting so many research attentions from 1950s until now. Scholars study it through different perspectives within various disciplines. This thesis summarizes several important perspectives in urban study in order to review the literature regarding waterfront development in the next section.

2.4 Research perspective for urban waterfront development

Waterfront development became a noticeable phenomenon in the 1950s and 1960s after the success of several projects. Normally, it is believed that it started in North America as a process of inner-city regeneration and then spread to many other parts of the world (B. Hoyle, 1995). Waterfront development has drawn academia interest and become a independent discipline since the 1960s (Marshall, 2001b). Nevertheless, compared with other topics in urban study, waterfront development is a relatively new discipline which is not as well-developed as other areas so far. This sectionwill summarize and review different research perspective of waterfront development.

2.4.1 The perspective of urban transformation

As mentioned in the beginning of this chapter, waterfronts carry cities' early memories and culture DNA. It is a meaningful approach to study the urban history through analyzing the evolution of waterfronts. For example, Han Meyer (1999a) uses a large number of maps, documents, pictures and graphics and elaborated on the history of waterfront development in several European and American cities. He selects four typical port-cities including London, Barcelona, New York, and Rotterdam to discuss their development processes, from the establishment of these ports and cities to the post-industrial period. Through his historical study, the evolution processes of waterfront and the relationship of ports and cities are clearly shown. Similarly, as it has shown in the beginning of this chapter, B. S. Hoyle, et al (1988) analyze the waterfront development through a historical perspective through the case study of Marseille, France, in which he concludes that waterfront development could bedivided into five common stages.

In the past few decades, urban transformation is a very important and widely-adopted approach to carry out the research of waterfront transformation. This approach encourages scholars to explore the past of waterfront, which could provide massive information for the present development. Through the comprehensive analysis of waterfront transformation, it might be understood that not only the process of waterfront development but also the dynamics and mechanisms. Generally, studies from this perspective are either based on chronological orders or historical events.

2.4.2 The perspective of relationship between waterfront (port) and city

It is almost impossible to conduct the study of waterfront development when it is isolated from the whole urban context. In decades, research embarking on the interrelationship of waterfront and city gained great achievements. To be more specific, studies can be further divided into two sub perspectives, namely the interrelationship of port and city and the connection of waterfront and other parts of the cities.

From the previous discussion, we understand that the evolution of ports and cities could be quite influential on each other. According to Hoyle (1988), especially from 1970s to 1980s, substantial change could be witnessed in the interrelationship between port and city on a variety of scales and dimensions. Brian Hoyle (2000) introduces two models to demonstrate interrelationship of port and city, namely the port-city interface model and the port-city evolution model. The first one reflects "a controversial port-city interface zone of conflict and only occasional collaboration" ; while the second one, on the other hand, could evoke a renewed collaboration as waterfronts are revitalized(B. Hoyle, 2000).

Since the development of waterfront became highly visible to the public, the relationship of waterfronts with other parts of cities also attracts research interests. Two issues of waterfront development are frequently mentioned, namely accessibility and connectivity. Waterfront developments are not isolated projects and the success of the waterfront revitalization lies on the connection between the waterfront and other urban districts (Greenberg, 1996). Richard Marshall (2001b) gives a comparative study about the connectivity of waterfront in Vancouver and Sydney. He concluded that the connection in Sydney is weak because of free highway system, zoning plan of waterfront and bad cooperation between authorities. On the contrary, the connectivity in Vancouver is enhanced through the reestablishment of a pedestrian network, adoption of an arterial system and so on.

Apparently, through this perspective, scholars could enhance their understanding of waterfront development as it is put in a large urban context. The development processes of waterfront (port) and city could have significant influence on each other. For over decades, the practice and research to solve inner-city decay and urban sprawl through waterfront revitalization are primarily based on this point of view.

2.4.3 The perspectiveof economic restructuring and social impacts

The social-economic perspective discusses the economic restructurings and its social impacts during waterfront construction and redevelopment. From comparative studies on waterfront redevelopment practices among developed countries, scholars identified a common phenomenon: the popular physical redevelopment approach in advanced post-industrial waterfront is in fact an economic restructuring process dominated by the transformation of capital and is also a process of spatial restructuring and social exclusion (Smith & Williams, 1986; Tweedale, 1988).

Studies examine the economic restructuring process through physical reconstruction and its relationship with the abandoned waterfront brown field. Using economic theory on capital accumulation and capital transformation, it explains how capital investment in the built environment has become dominant in a leading era of financial capital (Harvey, 1974; Tweedale, 1988). Analysis indicates that the economic restructuring of waterfront areas in many advanced countries in fact "transfer capital from the primary to the secondary circuit" (Tweedale, 1988) and adapt land uses to new economic activities (H. Carter, 1986). Studies go further to investigate the social impact of this economic restructuring, and indicate that it has ignored the social process in the redevelopment of waterfront areas and failed to address the problem of social injustice in terms of local people's access to employment, housing, education, etc.(Tweedale, 1988).

Using a socio-economic point of view, it develops the understanding of the way capitalism functions in the built environment and influences land use, and examines the cause and symptoms of social impacts. By addressing the problems of a partial understanding of physical and economic restructure and the total ignorance of social response in waterfront development, this approach contributes to the concern of social inequity.

2.4.4 Perspective of urban regeneration

Regeneration of waterfront provides an integral approach to the study of waterfront and "spiritual and physical renewal" for urban areas (Roberts, 2000). It seeks for a partnership management to regenerate a derelict urban place and includes multi-sectors like economy, housing, and transportation (Marshall, 2001a; Roberts, 2000). Urban regeneration aims to revitalize the economy, promote social well being and improve the physical and urban image in the long run (Roberts, 2000).

In the 1980s, there emerged a need for a comprehensive approach to study waterfront development, which could be identified as waterfront regeneration. The purpose of waterfront regeneration is to solve problems caused by a single economic restructuring or physical development by government-led top-down or private sector-led reconstruction and renewal (Meyer, 1999b). Meanwhile it combines physical and economic revitalization with many other fields, such as the social and environmental issues, to make strategy plans and decisions (Couch, 1990; Roberts, 2000).

The approach of waterfront regeneration pays attention to the cooperation of multiple disciplines and fields, such as accessibility, attractiveness, image making, economic investment, affordability, community participation, leisure, social life and tourism(B. Hoyle, 2000; Kemp, 2003; Millspaugh, 2001). As waterfront regeneration favors a partnership framework, it seeks the integration of factors (A. Carter, 2000). As Hoyle (B. Hoyle, 2000) concludes "the arrow becomes more sharply targeted as the planning process eventually reaches a practical and workable solution. ", even though wide topics are involved in waterfront regeneration, the urban planning process needs to formulate a strategy to implement.

Studies from this perspective investigate the integration of disciplines, conflicts and history(B. Hoyle, 2000). Waterfront development based on physical and economic aims has been criticized to be an isolated approach and has caused conflicts between government support, private capitalism and public goodwill, between economic growth and social isolation, between developer and neighborhoods, and so on and so forth. (Doucet, 2010; Kasintiz & Rosenberg, 1996; Seguchi & Malone, 1996). The regeneration approach tries to solve these conflicts by the integration of strategic implementations. Discussion can be found on comprehensive waterfront regeneration initiated by ideas like tourism, sports, recreation, festival, housing, public space, heritage conservation, natural resources conservation,etc.(Edwards, 1996; Gospodini, 2001; Marshall, 2001a; J. McCarthy, 1998; Millspaugh, 2001; Sofield & Sivan, 1994; Usavagovitwong & Posriprasert, 2006; Vallega, 2001).

Waterfront regeneration provides an approach for the issues of waterfront development. It emphasis the economic and physical aspects, and widens the understanding on urban waterfront development in social and environmental aspects. It requires cooperation, partnership that would contribute to a better planning process. The comparative studies of cases provide insights into the complicated process of decision making as well as the ways of implementation and management.

By following the examples and learning lessons, some mistakes can be avoided in practices elsewhere. However, as Roberts (2000) states, regeneration is a comprehensive system, it has different input, undertakes a different process, and will have a different output based on different contexts. A simple emulation of successful examples may not always lead to successful regeneration. Secondly, regeneration pursues long-term development rather than a short-term accomplishment. In practice, it is hard to measure the success of conflict reconciliation and balance of interest. In this case, many waterfront regenerations in practice are still inevitably weighting the economic gains prior to others' needs(B. Hoyle, 2000).

2.4.5 Perspective of urban design

According to Gosling (2003), urban design theory was recognized as a independent discipline in the 1950s and 1960s, against the background of unprecedented urban change in America. Almost at the same time, the study of waterfront development emerges under the same urban change background in United States, which is the inner-city decay and urban sprawl. In fact, in the beginning, the study of waterfront was inspired by the several urban design cases in urban waterfront with the intention to attract people back to downtown. Since then, waterfront study has always been related to urban design projects. A great number of publications could be found which address the study of waterfront development based upon the perspective of urban design as it is could be acknowledged as a primary one. It is impossible to give an overall review due to its complexity and so many components contained in urban design. Therefore, this thesis will select several aspects to discuss in the following part.

Due to the uniqueness and significance of waterfronts, in recent decades many scholars contributed to the urban design practices in waterfronts. These studies are normally carried out by project-based approaches in order to provide special design principles for waterfront developments. For instance, L. Azeo Torre (1989) selects 24 well known waterfront projects in the US to explain how to achieve successful waterfront development. According to him, mixed land use, water-related activities, effective management and financial feasibility are the key issues for the success of waterfront development. Using Baltimore Inner Harbor as a case study, Yang (2006) uncovers the theoretical themes might be the key to bring culture value in waterfront developments. William(2009) explores the possibilities of using thresholds to fulfill the mixed land use development for downtown waterfront. In general, these researches focused on specific urban design cases. Therefore, the knowledge provided by them can only be applied to limited regions.

In addition, a main characteristic of waterfront is its high visibility. What has been mentioned is that waterfronts in post-industrial cities are often redeveloped as public places. Being the particular public open space in urban context, issues regarding its publicity might be more sensitive than other urban sectors. The example shown in the previous section examines waterfronts' connectivity and accessibility through a comparative study of Vancouver's and Sydney's waterfronts (Marshall, 2001b). Similarly, Lim (2005) used Singapore River and Marina Bay as case studies and discussed the transformation of waterfront based on the point that "public space and public life is a result of manifestations on the waterfront". The study of public space is also involved in the research of social justice and public participation. For example, the importance of public participation in the process of waterfront development is emphasized by Wrenn(1983), Hoyle (1988) and Torre (1989). Apparently, studies from this perspective provide an effective way to extend the understanding of waterfront development in the larger urban context.

Besides, the networks of water bodies and waterfront shape the urban forms to some extent. The morphology of waterfront could be identified by their physical features and socio-cultural activities that take place in these areas. Morphology recently is adopted as an effective research methodology to understand waterfront development by many scholars. Lin-xue (1999) revealed the significance of waterfront morphological evolution in the process of site regeneration and outlined the design suggestions by the analysis of the elements of the waterfront. Similarly, Hu Wen Na (2004) discussed the common waterfront morphology, which is termed as "two-bank waterfront" in order to define the area along the river. Nonetheless, current research regarding waterfront morphology fails to address different urban forms' abilities to adapt to the climate change and water level change.

In summary, the similarity of these studies is that they are all based on specific urban design projects. Their objective is to enhance the design and implementation process of waterfront development. Because these studies are directly related to urban design projects, their advantages have strong practicality and pertinence. Nonetheless, project-based study also has limitations as it is only suitable for a specific locations or regions. Consequently, its contribution for the knowledge and research methodology seems to be inadequate. Moreover, climate change and sea level rise have already threatened waterfront developments in most of coastal cities. Therefore, there is a need to propose widely effective adaptations of sea level rise in order to complete urban design strategies for waterfront development.

2.4.6 Perspective of sustainable development

The concept of Sustainability or sustainable development was put forward in the 1987 report of the World Commission on Environment and Development, Our Common Future. It states that sustainable development is "to meet the needs of the present without compromising the ability of future generations to meet their own needs (Brundtland, 1987, p. 8)".

Especially in this decade, climate change and rising sea level has raised new challenges for the waterfront in coastal cities development as previously discussed. Under this circumstance, many scholars dedicated themselves to waterfront research in regard with sustainable development. Among this research, three aspects are discussed here, namely urban hydrology and waterfront engineering, integrated coastal zone management, and resilience development and adaptation research.

Urban hydrology is a science investigating the hydrological cycle and its change, water regime and quality within the urbanized zones. Urban hydrology is a link in a number of sciences dealing with the problems of ecology, environmental protection, conservation and rational use of the water resources (Viessman, Lewis, & Knapp, 1977). The research interests include the supervision and treatment of urban water quality, preservation and management of urban water resource, influence of urbanization on hydrology and flood protection. Generally, this approach would use scientific technologies to carry out studies, such as computational fluid dynamics (CFD) and geographic information system (GIS). Together with the waterfront engineering, the contributions of this approach for the sustainable waterfront development could not be underestimated.

Integrated coastal zone management (ICZM) is a process for the management of the coast resources using an integrated approach, regarding all aspects of the coastal zone, including geographical and political boundaries, in an attempt to achieve sustainability (Cicin-Sain, 1993). This concept was born in 1992 during the Earth Summit of Rio de Janeiro. ICZM is a dynamic, multidisciplinary and iterative process to promote sustainable management of coastal zones. It covers the full cycle of information collection, planning (in its broadest sense), decision making, management and monitoring of implementation (Cai, Su, Liu, Li, & Lei, 2009).

Obviously, for coastal cities, waterfront development is an essential component in urban planning. In the 21st century, with the rising sea level and frequent oceanic disasters, this approach becomes even more important. Nevertheless, up to now, the main concerns of ICZM studies are the natural elements so that the contribution from urban planning and urban design are always neglected or disregarded

As emphasized in this thesis, climate change and sea level rise might challenge the way which we adopt to build our waterfront in coastal cities. In order to respond, urban planners and researchers make efforts to enhance the suitability in waterfront development. Their goal is to integrate the waterfront development and climate change, to update our waterfronts in order to make them more resilient and to provide effective and feasible adaptations.

In regard to the seriousness and urgency of climate change impacts, remarkably high attentions have been given to the research of adaptations, with the intention to promote sustainability and resiliency of waterfronts. Walsh et al. (2004), Davoudi, Crawford, and Mehmood(2009), Wilson and Piper (2010) are some of the well-known scholarly works on waterfront research. Especially, Bush, et, al.'s (1996) book discuss innovative hypotheses regarding urban fabric and spatial arrangement might have impacts on waterfronts' resiliency. This thesis will examine these studies in the following chapters, focusing on their contributions and limitations.

2.5 Chapter summary

The discussion above provides a general picture for recent studies of waterfront development. Based on this literature review, it could be concluded that waterfront is a quite essential sector in urban study. All around the world, local governments are or will be motivated to promote waterfront developments in order to demonstrate their ambitions, upgrade urban environments, and improve the quality of urban life. However, the challenges from climate change and sea level rise are making the process more complicated than ever before. New requirements have been put forward, when people are not ready to change as most of current projects do not concern very much about the threats from changing climates and rising seas.

In this chapter, main perspectives of waterfront study in recent decades have been introduced, which are mainly primary and widely adopted ones. Generally, studies might not discuss topics based on sole perspectives but combine two or more. Under the circumstance of being threatening by climate change, research regarding waterfronts should shift the focus in order to arouse people's attention and provide solutions. In my opinion, in order to enhance waterfront resiliency and provide adaptations for urban design practice, it might be the most significant research approach to combine the perspectives of urban design and sustainability.

For coastal cities in low land, the safety of the waterfront development becomes the first priority. It is important to understand how urban planning systems respond to these in the calamitous future. The appropriate layout of the coastal area and necessary adaptations to the sea level rise and extreme events should become the focus of the waterfront development in the future. Endangered coastal areas might have to rethink the ways to build waterfronts. Regarding this topic, following chapters will expend discussions with more information.

Chapter 3 Adaptations to extreme sea level events

In the last chapter, it is concluded that serious damages seem inevitable if coastal cities take no action to respond to sea level rise and other catastrophic events related to climate change. Apparently, climate change has become the most challenging threat to human society. In recent years, great effort was invested into the study of climate change. In this chapter, with the focus on adaptation to extreme sea level events, literature regarding response strategies will be reviewed from the perspective of urban researchers. Based on the literature review of adaptation, the research hypotheses will be stated.

3.1 Adaptation versus mitigation

Broadly speaking, response strategies to cope with climate change can be categorized into two main sections, namely: (1) mitigation, and (2) adaptation (Sharma & Tomar, 2010). In recent decades, the debate about which is the most appropriate/effective response strategy to cope with climate change between adaptation and mitigation has attracted a great deal of attention. IPCC (2007) defines mitigation as an anthropogenic intervention to reduce the sources or enhance the sinks[1] of greenhouse gases, while Sharma and Tomar(2010) refer to adaptation as the capability to "adjust to climate change (including climate variability and extremes), to moderate potential damage, to take advantage of opportunities, or to cope with the consequences" (p.452). Among these two strategies, Davoudi, et al.(2009) advise that mitigation should have the priority since it is the primary form of adaptation. Accordingly, mitigation might be more important and effective than adaptation. Generally, Davoudi's theory is well accepted over the past few decades. Compared to adaptation, most counties and organizations would give more attention to mitigation. Biesbroek et al. (2010) argue that for the last two decades, climate policies formatted by European countries have focused almost exclusively on mitigations, such as setting targets on reducing carbon footprint. However, this point of view has been recently questioned.

---

[1] Sinks means carbon sink, which refer to artificial or natural reservoir that stores carbon-containing chemical compound, for more information please refer to Fan et al.(1998), Pacala et al. (2001), and Pan et al.(2011).

After the turn of the century, with increasing impacts of climate change being observed, adaptation is now emphasized in the policy agendas (Biesbroek et al., 2010). According to Reid and Huq(2007), even if society puts an immediate halt to GHG emissions, the average temperature could continue to rise for decades due to the long life of carbon dioxide in the atmosphere, which could exceed 100 years. Accordingly, Huq et al. (2007) argue that "adaptation to climate change is equally important", especially for vulnerable cities. It is believed that most of these vulnerable cities are not big emitters of GHC—some of them are situated in developing countries, which are highly threatened by climate change-related extreme events (Dossou & Glehouenou-Dossou, 2007).

There is currently little research that examines how adaptation and mitigation affect each other (Sharma & Tomar, 2010). It is clear enough that adaptation and mitigation are equally important measures that address climate change. Mitigation might have global and long-term benefits, whereas adaptation could offer immediate and local benefits (Sharma & Tomar, 2010). In the scope of response strategies to extreme sea level events, adaptation tackles the notion of urgency more than mitigation since sea level rise might not be stopped or slowed down in the next several decades. In fact, for certain vulnerable areas, some adaptive measures must be carried out immediately while long-term and global targets should not be compromised in the process. In other words, the possible influence on mitigation must be carefully assessed and considered when the local adaptive strategies are being carried out.

3.2 The concept of adaptation and its evolution

3.2.1 The changing meaning of adaptation in urbanism

Almost all English dictionaries give the similar meanings to "adaptation". For example, in the *Longman Dictionary of Contemporary English*(Plachouras & Ounis, 2007), it is defined as "the process of changing something to make it suitable for a new situation" (p.17). Currently, the word "adaptation" often relates itself with the issue of climate change and in architecture and urbanism adaptation is, in fact, not a new concept.

The concept of adaptation design or design for adaptability was first developed by a group of architects in order to respond to the changes of life and daily scale of inhabitants through time (Ruskeepää, 2011). In the 1950s to 1970s, a group of architects (many of whom were members in the famous architects alliance "Team 10") devoted themselves to housing design, with the intention to improve "expandability, versatility, convertibility and fluidity" (Ruskeepää, 2011). After World War II, a mass-based and standardized housing construction program was put in operation in order to solve housing shortage in Europe. The standardized design for housing tended to be monotonous and incapable to adapt the change of post-war social life. According to Risselada and van den Heuvel(2005), the concept of adaptation and adaptability might have been initially proposed by Team 10 to advocate against the monotonous and standardized solution for housing program. Subsequently, the concept was also adopted into larger scales, such as neighborhoods, districts and cities.

The idea of adaptability in architecture and urbanism has been developed to be a capacity to adapt to changing circumstances, including socio-cultural, economic and environmental changes (Ruskeepää, 2011). In recent decades, climate change and global warming are challenging our approach of designing, building, and managing our cities. In his seminal book, Lynas(2008) summarizes a large number of the impacts of global warming and categorizes them into six chapters according to increments in global surface temperature (from 1 degree to 6 degrees) which are triggered by heat wave, heat island effect, flood, drought, sea level rise, famine, desertification, and so on. Although there are still debates on the certainty of climate change, the author points out that our generation is presented with a critical responsibility to shape the fate of humanity. Under this circumstance, it is important that more attention be given to climate change when adaptation activities are discussed.

According to IPCC, adaptation to climate change refers to "the adjustment in natural or human systems in response to actual or expected climates and weather, or their effects, in order to moderated harm or exploit beneficial opportunities" (Parry, 2007). IPCC divides adaptation into three categories, including anticipatory adaptation (proactive adaptation), autonomous adaptation (spontaneous adaptation), and planned adaptation (ibid, p.869).

- Anticipatory adaptation, adaptation that takes place before impacts of climate change is observed. (Also referred to as proactive adaptation.)

- Autonomous adaptation, adaptation that does not constitute a conscious response to climate stimuli but is triggered by ecological changes in natural systems and by market or welfare changes in human systems. (Also referred to as spontaneous adaptation.)

- Planned adaptation, adaptation that is the result of a deliberate policy decision, based on an awareness that conditions have change or is about to change and that action is required to return to, maintain, or achieve a desired state.

Based on this definition and classification, Cohen (N. Cohen, 2011) summarizes that climate change adaptation may include "forecasting impacts, monitoring current conditions, risk assessment and education, mapping vulnerable places, disaster preparedness, liability assessment, promoting behavioral change (e.g., water and energy efficiency or decreased automobile use), building adaptive capacity, and retrofitting built environments (e.g., installing building insulation, photovoltaics, and green roofs, or planting more street trees) " (p.4) . Cohen also emphasizes that new technologies play a great role in adaptive measures but highlights that the cost of such technologies can make the testing and implementation of adaptation strategies expensive.

In fact, taking measures to adapt to climate variability is not a new concept. According to Schiffman (2011), humans have a long history of adapting to climate change, including managing resources, irrigating crops, constructing reservoirs and other actions taken by individuals, communities and governments. Wu (1995) summarizes some projects done in China during the period of antiquity when the Chinese faced threats from floods. Among a large number of cases introduced, many methods and principles are still valid in modern urban developments. Similarly, Ruskeepää(2011) also introduces some protective measuresin ancient Dutch cities, which were designed for floods and are still functional now, including some dikes and sea walls.

As it is emphasized in this thesis, climate change might be the most important issue in this century since it has already cause enormous impacts on human society. Ruskeepää (2011) addresses that climate change has become the most prominent impetus for a re-thinking of adaptation in urbanism now as well as the alteration in the physical dimension of built environment. Nowadays, adaptation is a frequently mentioned term in the study of climate change as it seeks the best way to modify cites to cope with the expected changes based on the projected possible impacts (N. Cohen, 2011).

3.2.2 Bottom-up and top-down adaptation

Looking at things from a different perspective, Wilby and Dessai(Wilby & Dessai, 2010) differentiated adaptation to climate change and put them into two main broad categories, namely top-down adaptation (also known as 'scenario-led') and bottom-up adaptation. The former involves local climate projections based on gas emission scenarios. The latter, on the other hand, focuses on "reducing vulnerability to past and present climate variability", typically applied after extreme disasters and climate change events (Ibid, p.181). The authors explain that the term of "bottom-up" is used to describe adaptation measures that are initiated at the level of individuals, households and commodities. In other words, top-down approach is characterized by anticipatory and proactive adaptation while bottom-up approach is spontaneous adaptation and planned adaptation (Parry, 2007).

It is argued that although top-down approach is the more widely represented of the two, and research groups have provided quite a lot scientific evidence to predict the future climate change, there are very few tangible examples of anticipatory and planned decisions (Wilby & Dessai, 2010). Dessai and Hulme(2004) argue that, currently, the majority of research studies stop at the stage of impact assessment and top-down approach might be highly dependent upon the results of global climate models (GCMs) and experimental design, which may also contain a large extent of ambiguitiesand uncertainties (Dessai & Hulme, 2004).

However, in practice, other factors including economy, social equality, education status, physical and institutional infrastructure, and access to natural resources and technology can also help to determine local climate vulnerability (Brooks et al., 2005). Adaptation strategies focusing on such factors might be classified into the category of bottom-up approach. Apparently, the success of adaptation policies might rely highly on such factors. Therefore, combining these two approaches is the only way to formulate effective adaptation strategies.

3.2.3 Adaptation across scales

In the last section, it is already mentioned that the impacts of climate change might affect us at different levels, from the scale of a country to that of a single individual. For instance, some impacts might be a countrywide or even regional issue, such as food scarcity, famine, and widespread epidemic, while others might have an influence on a metropolitan or city area, such as floods and hurricanes. Meanwhile, other forms of impacts might affect some districts or neighborhoods in vulnerable locations. In other words, impacts of climate change may cross scales.

Therefore, adaptations should have the capability to work at different spatial and societal scales in both physical and ecological systems as well as in human adjustments to resource availability and risk. Adger et al. (2005) argue that it is necessary that the sustainability of adaptations is evaluated against different criteria at different scales. To achieve this, adaptation is required to involve participants at different levels, including individuals, groups, organizations, and local and central governments.

In the past two decades, many countries have formulated official documents to address the issue of adaptation to climate change. For example, in April 2007, Department of Climate Change and Energy Efficiency of Australia published the "National Climate Change Adaptation Framework" in order to guide the development of climate change policies nationwide. For the same purpose, in July 2013, UK government release the "The National Adaptation Programme: Making the country resilient to a changing climate". Similarly, the National Climate Change Secretariat (NCCS) of Singapore also drew out a document in 2013, entitled "Adaptation Measures", on how to deal with the threats of climate change.

Generally, compared to broad guidelines, specific measures of certain issues are much more difficult to address. It has been proven that more effort and resources should be given to smaller-scaled entities, such as communities and neighborhoods (Osbahr, Twyman, Neil Adger, & Thomas, 2008). As Adger et al. (2005) mention, adaptation to climate change could involve both building adaptive capability and implementing adaptation decisions. Sometimes, actions that increase the adaptive capacity at the smaller scales may be individual behaviors, but it is important to transform these actions into implementable policies. Meanwhile, quite a lot of research has shown that the density of cross-scale interaction is the key factor to successful adaptations, especially for local resource management regimes (Berkes, 2006; Cash et al., 2006).

3.2.4 Incremental adaptation and transformation adaptation

Following this conception, Kates et al. (2012) creatively divide adaptation into two broad categories, namely transformational adaptation and incremental adaptation. According to them, incremental adaptation refers to "extensions of actions and behaviors that already reduce the losses or enhance the benefits of natural variations in climate and extreme events" (p.7156); on the other hand, three classes of adaptations could be described as transformational, including "those that are adopted at a much larger scale or intensity, those that are truly new to a particular region or resource system, and those that transform places and shift locations" (Kates et al., 2012). They also state that the majority of the planned or implemented climate change adaptations may belong to incremental ones. For instance, the US National Research Council's Panel on Adapting to Impacts of Climate Change (2010) specifically lists 314 adaptations to certain impacts of climate change, but Kates et al. (2012) argue that only 16 (5%) of the 314 measures could be classified into innovative adaptations, which had never been applied in the United States. The lack of motivation to develop transformational adaptations could be attributed to multiple reasons, which will be discussed in the following part. However, as Schiffman (2011) mentioned, past experiences of handling climate variations might not be adequate for future climate change events. Therefore, the lack of transformational adaptation could result in the rise of damage since the observed impacts of climate change increase in terms of frequency and intensity.

All in all, extreme climate change events in future years might become the incentives for governments and research groups to devote more effort and resources to development more innovative adaptations which could arm vulnerable locations and enhance their resiliency.

3.2.5 Section summary

In summary, although human beings have been adapting to climate variations for time immemorial, climate change in this era is arguably more serious than it has been in any other period in human history. It has become a force and drives human to make changes. However, both the research and actions of adaptation to climate change are not sufficient yet, especially for transformational adaptations, which may be due to several difficulties, including the uncertainty of climate change, expensive cost and behavior barriers. These difficulties will be discussed in the following sections.

3.3 Adaptation is difficult to implement

Although it is argued that for some highly vulnerable locations, adaptation would appear to be a more urgent response compared to mitigation policies, adaptation strategies and the success rate of these adaptive measures are in fact more complicated. Regarding the success of adaptation to address climate change, Adger et al (2005) believe that effectiveness, efficiency, equity and legitimacy are the most essential factors. Similarly, Biesbroek et al (Biesbroek et al., 2010) also develop a system to assess adaptation policies at the national level, using the assessment tool to compare the national adaptation strategies (NAS) of six countries in Europe. According to their evaluation, there are few examples that are identified to be effective. In fact, presently, there are certain barriers that make adaptation strategies difficult to be formulated or implemented, which mainly include uncertainty, cost, and institutional and behavioral barriers.

3.3.1 Uncertainty of climate change

According to the Symposium for Policymakers in IPCC Fifth Assessment Report (Stocker et al., 2013), the projections of the global mean surface temperature change and global mean sea level rise are based on different scenarios and the results vary greatly. Currently, there is no disagreement on the view that earth is getting warmer but the debate continues in academia regarding the projected incremental rise of the global mean temperature. As mentioned earlier, according to Lynas' (2008) summary, the predictions for the incremental rise of global surface temperature at the end of this century range from 2 to 6 degrees. Meanwhile, the reasons for the cause of climate change are also controversial.

When the uncertainty of climate change and the huge number of potentially vulnerable populations cannot be accurately estimated, formulating adaptation strategies might be a sensitive issue for policymakers (Wilby & Dessai, 2010). Kates et al. (2012) also address that, with the exception of earth's most vulnerable places (e.g. very low-lying islands or deltas), both the impacts of extreme climate-related events and the benefits of adaptations, especially transformational adaptations, remain unknown.

Presently, research assumptions about the possible impacts caused by climate change events and the efficiency of adaptations are based on two methodologies, namely: (1) observation of similar events, and (2) simulation based upon projected scenarios. For example, Wong (1992) stated the impacts of one-meter rise of Singapore's sea level based on the observation data of a high tide event in 1974.

The results and conclusion could provide an outlook of the future, but no one could validate these findings with certainty.

Therefore, considering these uncertain factors, public policymakers might hesitate to formulate and implement adaptations for climate change; thus, it could be concluded that uncertainty will certainly undermine their effectiveness.

3.3.2 Cost

Carrying out metropolitan-level and regional-level adaptation measures could incur huge expenditures, including money, time, and resources. For example, a study carried out by Ng and Mendelsohn (2005) in Singapore shows that, based on the projection of future sea level rise, the annual cost of protecting the coastal areas will rise from US$0.3–5.7 million to US$0.9–16.8 million between the years 2050 and 2100. However, the authors also state that although the cost to adapt to sea level rise is significant, the irreversible loss of land to inundation would far exceed the expenditures incurred to prevent this loss. Similarly, Heberger et al. (2009) show that if the sea level rises one meter in California, US$100 billion of property will be threatened and protective measure might cost US$14 billion and 10% annual operation and maintenance costs. According to estimates by the Environment Agency (2009), in order to protect the Thames estuary, the transformational option may cost £4.2 billion. The Netherlands also proposes to fund an integrated and ambitious program, starting in the year 2020, with the intention to protect the nation from floods, drought, and coastal protection, which will spend 1 billion Euros per year (Commission, 2008).

Additionally, some impacts might be exacerbated by numerous urban development practices, such as development on floodplains and coastlines which may increase vulnerability to flooding and coastal erosion and make adaptation more difficult (N. Cohen, 2011). Therefore, some adaptations require the downscaling of development in vulnerable areas. For example, both Bush et al. (1996) and Davoudi et al. (2009) emphasize the importance of restricting the development density in vulnerable coastal areas. Scaling back developments might result in losses of land value, which some governments might also consider as a cost.

Based on the arguments above, it is evident that most adaptation measures may be costly; especially those that are intended to address the large scale impacts of climate change. For other more autonomous and small-scale adaptation strategies, the individual costs may be low, but the collective expenditure cannot be underestimated. Moreover, Kates et al. (2012) emphasize that the costs of transformational and novel adaptations might be higher than incremental and familiar ones.

3.3.3 Institutional and behavioral barriers

In addition to uncertainties and costs, another considerable obstacle are institutional and behavioral barriers. Kates et al. (2012) state that barriers to adaptation, especially for transformational adaptation, include "legal, social, and institutional conceptions of rights, longstanding resource allocation policies, customary protection and entitlements, and behavioral norms" (p.7158).

At same time, Cohen (2011) raises the issues of equity and justice in the process of carrying out adaptation. He believes that insufficient citizen involvement and inadequate efforts to promote adaptive ability among marginalized and vulnerable populations might undermine the efficiency of adaptation. The problem of limited involvement might be due to the scarcity of public awareness and engagement. After a comprehensive analysis of Japanese newspaper reports from January 1998 to July 2007, Sampei and Aoyagi-Usui(2009) reveal that, before January 2007, coverage of global warming had "an immediate but short-term influence on public concern". They suggest the more effective communication of climate change and related strategies is required. Lorenzoni et al.(2007) conducted a survey showing that the weak public concern about climate change and limited behavioral response may discourage the target of climate policy in the UK.

As mentioned above, adaptation to climate change requires multi-level cooperation. Therefore, effective communication and collaboration across different levels and different government agencies would be essential, and new improved processes would need to be put in place in order to achieve successful adaptation (Neil Adger et al., 2005; Osbahr et al., 2008).

In summary, adaptation to climate change is difficult to formulate and implement. It requires huge investments and involvement of time, money and resources. More importantly, in order to respond to climate change, people have to change their mindsets—in other words, rethinking approaches to building and managing resources, environments and cities.

3.4 Adaptations to sea level rise and extreme sea level events

As illustrated by the discussion in the previous section, it is quite clear that future sea level rise would have significant impacts on urban development and urban planning systems must take early response measures before the huge damages are done. Generally, Dronkers et al. (1990) label adaptation to sea level rise and extreme sea level events by three main approaches, namely: (1) retreat, (2) protection, and (3) accommodation. Before taking any actions, however, it is necessary to carry out coastal risk assessment.

3.4.1 Coastal risk assessment

According to the Strategies for Adaption to Sea Level Rise (Dronkers et al., 1990), there are three objectives of coastal risk assessment, which are to "avoid development in areas that are vulnerable to inundation, to ensure that critical natural systems continue to function, and to protect human lives, essential properties and economic activities against the ravages of the seas" (p.6).

Nowadays, the effective way to conduct coastal risk assessment is through risk mapping. In their informative book, Bush et al. (David M. Bush et al., 1996) elaborates the method of coastal risk mapping as "to map the level of relative property damage potential for the coast and assists community officials to reduce impacts from natural disasters and allows individual property owners to choose sites and purchase property in a more knowledgeable way" (p.41). Concerning the complexity caused by massive factors, the authors propose a useful methodology to identify the risk categories of the location. These factors can be organized into four main categories: (1) coastal settings, (2) situation of development, (3) shoreline engineering type and location, and (4) storm response capabilities. Moreover, the coastal risk management programs usually include governmental controls, private sector incentives, and public support. In addition, successful coastal risk management programmers require public education and need to gain broad based support and public participation to ensure equal representation of interests (David M. Bush et al., 1996).

In summary, risk assessment and management are essential for coastal areas when dealing with rising sea level. Accurate risk mapping is fundamental to guide adaptation policies.

3.4.2 Retreat

Retreat means abandonment of land and structures in vulnerable areas and re-settlement of inhabitants (Dronkers et al., 1990). To be specific, retreat strategy is composed of three policies, including: "(1) preventing development in areas near the coast; (2) Allowing development to take place on the condition that it will abandoned if necessary (planned phase out); (3) No direct government role other than through withdrawal of subsidies and provision of information about associated risks" (p.6).

However, for small island states, such as Singapore and Hong Kong, retreat is not an acceptable option. There would be little land for resettlement. Meanwhile the economic activities would be seriously downgraded. Ng and Mendelsohn (2005) conduct a study to calculate the potential economic costs caused by sea level rise in Singapore. The cost of protection and the cost of inundation are compared in this study. It is found that the cost of retreat is extremely high while protection is the most desirable and efficient solution to sea level rise for the market land in Singapore. Even if construction and maintenance costs are higher than expected, the total protection cost is still significantly lower than the benefit (W. S. Ng & Mendelsohn, 2005). In table 3.1, the costs of inundation and protection are compared based on the projections of sea level rise in different scenarios by 2050 and 2010.

Total economic impacts on Singapore, from 2000 to 2100, for three sea level rise scenarios.   Table 3.1

2050	0.2m	0.49m	0.86m
Potential land inundated	2.23	5.45	9.56
Decadal value of land	1,872 (1,040)	4,539 (2,521)	7,980 (4,433)
Decadal protection cost	5.58 (3.10)	33.49 (18.61)	103.17 (57.32)
2100	0.2 m	0.49 m	0.86 m
Potential land inundated	3.96	9.70	17.02
Decadal value of land	6,769 (3,760)	16,599 (9,221)	29,090 (16,161)
Decadal protection cost	16.40 (9.11)	98.42 (54.68)	303.19 (168.45)
Present value of protection costs	0.30 (0.17)	1.80 (1.00)	5.55 (3.08)

Notes: The areas of potential land inundation are presented in Sq km. The values of land and related discounted values are in millions of 2000SG$ with millions of 2000 US$ in parentheses.

Source: W. S. Ng & Mendelsohn (2006)p. 211.

3.4.3 Protection

Protection strategies involve defensive measures and other activities to protect areas against inundation, tidal flooding, effects of waves on infrastructure, shore erosion, salinity intrusion, and the loss of natural resources (Dronkers et al., 1990). These adaptive measures are usually some engineering projects at off-shore and shoreline locations and can be put into two main categories: (1) soft protection, and (2) hard protection. Table 3.2 summarizes the most popular protection options.

Nowadays, protection, particularly hard protection, is the first choice amongst most coastal cities. For instance, Singapore, given a location surrounded by sea water, completed the construction of the S$226 million Marina Barrage in 2007 at the mouth of the Marina Channel. The unique 3-in-1 Marina Barrage project will not only help to increase Singapore's water supply but also alleviate flooding (Figure 3.1).

Current popular protection options — Table 3.2

Soft protection	Adding sand to beach	Beach replenishment/nourishment
		Beach bulldozing/scraping
	Increasing sand dune volume	Sand fencing
		Raise frontal dune elevation
		Plug dune gaps
	Vegetation	Stabilize dunes
		Plant marsh
Hard protection	Shore parallel	Seawall
		Bulkheads
		Revetment
		Offshore breakwaters
	Shore perpendicular	Groins
		Jetties
	Between the sea and estuary	Barrages

However, these protections are not panaceas. It is believed that some of them are no longer the environmentally-sensitive solutions that they are often made out to be, especially in terms of hard protection. For example, the seawall or the dikes could accelerate land erosion over time (Kelletat, 1992). Anderson (2007)

Figure 3.1 Marina Barrage of Singapore
Source:http://www.kohbrothers.com/projects

also posits that longshore currents might undermine the coastal ecosystem, which means that when sea waves are not exactly perpendicular to the beach, the waves would force the currents to move along the beach and take away the sand and if the waves encounter some type of barrier, such as the seawall, the longshore currents would grow in strength. He conducted a case study of the Bolivar Peninsula shoreline to demonstrate that building the seawall could force the longshore current to wash the sand out to another location. As a result, the land retreat rate was accelerated due to the loss of sand on the beach.

Compared to the hard protections, soft protections are more environment-friendly. Nevertheless, most of them are very expensive and need a long term operation. Beach nourishment is an effective way to slow down the rate of coastal retreat, which requires adding more sand on the beach (Dean, 2002). However, such added sands are not stable and the beaches must be nourished again and again. Moreover, in most the cases, the sands must be purchased from other cities or even imported from other countries (J. B. Anderson, 2007). Therefore, it does not present a sustainable and finite solution.

In summary, there is no one-stop-solution. To protect the waterfront, local governments have to combine several protection measures and integrated them together concertedly. Additionally, the protection engineering projects must reduce their impact on the natural system so as to allow the natural system to collaborate with man-made constructions. For a long time, the human perspective of coastal hazards tends to be biased, focused on the sea and the shoreline. Now, it is timely and pertinent that this perspective be changed or else revisited.

3.4.4 Accommodation

Accommodation means taking measures on buildings and other built environments in order to keep them operating normally under circumstances of sea level rise, and it requires advanced planning (Dronkers et al., 1990). Currently, in this category, the majority of research attention has been given to building sectors, such as the enhancement of both building codes and construction regulations (David M. Bush et al., 1996; Simin Davoudi et al., 2009). As architecture design and building technology are beyond the scope of this research, adaptations from this perspective will not be discussed in depth.

Case studies on a larger scale, including using projected sea level rise for master planning, land use planning, and evacuation route planning, can be found in literatures. For example, in the Netherlands, an advisory commission installed by the government, the Deltacommissie, published a report on the protection of the country against flooding in the coming century in September 2008. To reduce the uncertainty of projection of future sea level rise, a 130cm sea level rise by the year 2100 was set, which is far more than the estimates made by the IPCC in 2007 (Simin Davoudi et al., 2009). In Victoria, Australia, the Victorian Coastal Strategy 2002 encourages a program of vulnerability assessment within a 100-year planning horizon as a basis for detailed statutory planning. The Victorian Coastal Strategy is required to be taken into account by planning authorities through planning schemes. A range of studies and modeling assessments are underway to develop improved estimates for detailed planning (Simin Davoudi et al., 2009). Walsh et al. (2004) address that in coastal cities, sea level rise and related events need to be taken into account because their effects may turn apparent before 70 years, which is the typical replacement time of infrastructure and buildings.However, a research gap may exist on the meso-scale.

3.4.5 Meso-scale urban form-based adaptations

To date, only a few researchers have aimed specifically to propose the accommodation strategies in the perspective of urban form, urban fabric, and spatial planning. In 1996, Bush et al. (1996) published their informative and enlightening book, *Living by the Rules of the Sea*, in which the functions of urban road system and spatial planning are proposed in the response strategies. According to their theory, waterfront resilience could be enhanced through special urban design and spatial organizations. Nonetheless, only hypotheses could be found in their research. In other words, although the authors advocate such kind of adaptation, they provide no scientific proof to address the effectiveness. Almost 20 years later, systematic and comprehensive studies on this subject still cannot be found although some researchers mention the work of Bush et al. (1996) in a hypothetical manner (Simin Davoudi et al., 2009; Hurlimann et al., 2014; Vasey-Ellis, 2009; Wilson & Piper, 2010). For the purpose of formulating the research hypotheses, this study will summarize some urban form-based and meso-scale adaptations proposed by previous research.

1) Controlling the development density
Bush et al. (1996) emphasize the importance to restrict the development density in coastal areas. According to Davoudi et al., (2009) floods occur because urban soils cannot absorb or discharge fast enough the excess water that results from heavy rains or overflowed rivers and seas. Dunne (1984) suggests that the most obvious method of reducing runoff is to maintain as much as possible of the natural vegetation and permeable soils. Therefore, in vulnerable waterfronts, the development density should be strictly restricted (Hurlimann et al., 2014). In more dense and compact urban areas, streets may flood more easily and runoff drainage may be retarded, possibly resulting in deep flooding (Simin Davoudi et al., 2009). However, this issue is complicated by the high market value of waterfront land. In most cases, urban planners are confronted with the dilemma to balance coastal resiliency and land utilization efficiency. Although researchers repeatedly address the importance of restricting the development intensity in vulnerable waterfronts, academic or practical methodologies to formulate appropriate density are hardly found.

2) Orientation and placement of coastal road system
Additionaly, Bush et al. (1996) believe that: "If the direct line created by roads perpendicular to the shore could be interrupted, the amount of damage done by overwash and storm-surge flood and ebb waters could be reduced" (p.111). Therefore the solution is to avoid shore-perpendicular roads and four ways are introduced in their book to change the road system in order to mitigate the damage caused by floods, including: "(1) building a small "bump" on oceanward terminus of the shore-perpendicular roads; (2) adding a few simple curves in road; (3) building roads at some angle to the shoreline; and (4) interrupting the grid pattern so as to eliminate the conduit effect" (Bush et al., 1996. p.112).

Volusia County, Florida, USA (Figure 3.2), is discussed as an example, in which the "east-west roads parallel township……and thus at some angle to the shoreline" (Bush et al., 1996. p.112). Another example is Bogue Banks, North Carolina (Figure 3.3); the Land's End subdivision near the western end has a deeply curved road layout. Bush et al. (1996) argue that although these areas were likely built due to aesthetic reasons, it will certainly help mitigate property damage in storms.

Figure 3.2　East-west roads in Volusia Country, Florida, US
Source: Google.

Figure 3.3　Land's End subdivision near the western end of Bogue Banks
Source: Google.

Figure 3.4 shows a case study of Folly Beach, South Carolina, USA, to demonstrate the options of a waterfront road system, in which three plans are designed to reduce the impact of storm surge through replacement of the perpendicular roads with oblique ones (David M. Bush et al., 1996).The three plans use approximately the same amount of sand for dunes in original crossroads, but use increasing amount of street reorientation from a to c. Based on the same urban context, they do another modification on the road system through interrupting seaward roads and elaborating that the traffic might still work (Figure 3.5). They believe that the less shore-perpendicular roads, the more capable of the coastal area to respond to storm surge floods.

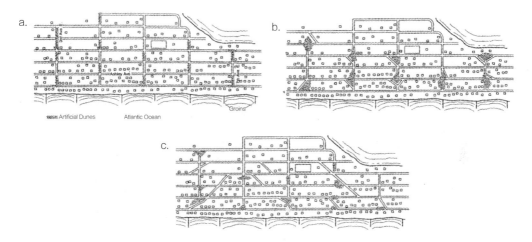

Figure 3.4   Three plans to reduce the impact of storm on Folly Beach
Source: Bush, et al, (1996) p.119.

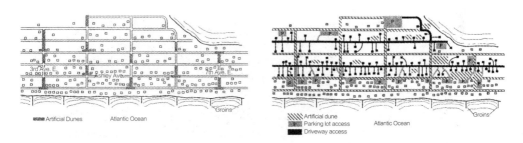

Figure 3.5   Modification of waterfront road system through interrupting seaward roads
Source: Bush, et al, (1996) p.120.

3) U-shaped building and Finger canals
Besides, Park et al. (2013) mention that the simulation of tsunami shows U-shaped buildings can amplify and concentrate the wave, which may upscale the impacts. However, in order to create more waterscapes, the U-shaped building is a popular form in waterfronts, especially for hotels and restaurants; and the scale of buildings is normally quiet large. Park et al. (2013) give an example for such urban

form, which is in Seaside, Oregon, US as shown in Figure 3.6. Due to the spatial characteristic of U-shaped form, waves might easily enter courtyards and is hardly dredged out. Therefore, the impacts caused by waves might be amplified. Similarly, Bush et al., (1996) also address this issue in their book in terms of residence design. With the example Myrtle Beach, South Carolina, US, they point out man-made finger canal in waterfront neighborhoods may cause same problems during ESLEs. Although U-shaped buildings, blocks and topographies are criticized, currently, no research has been found in existing literature to address how to deal with this issue and modify such forms in order to mitigate the impacts from ESLEs.

Figure 3.6 U-shaped building in Seaside, Oregon, US
Source: Google.

Figure 3.7 Finger canals in Myrtle Beach, South Carolina, US
Source: Google.

3.5 Research hypotheses

From a planner's perspective, the studies discussed above state that urban form and spatial layouts can play a positive role in responding to ESLEs. They provide innovative thoughts for the meso-scale urban form-based adaptation strategies. Nonetheless, the research does not yet give the methodology to prove the effectiveness of proposed adaptations. Therefore, their theories can only be considered as hypotheses, which need further testing Meanwhile, proposals, such as replacing the grid patterns by curved roads, might not be applicable mainly because theylacks strong justifications for hindering travel patterns by slowing down traffic on an everyday basis in order to prepare for infrequent extreme events.

To some extent, it remains unclear, particularly in the discipline of urban study, how urban form, the shapes and sizes of urban blocks, the combinations of buildings and open space, and other physical features can help waterfront development adapt to sea level rise and extreme sea level events. Based on above discussion and resent studies, the research hypotheses of this thesis are proposed as follows:

3.5.1 Hypothesis 1: block typology

- There are certain urban block typologies which can mitigate the impacts from ESLEs.

With the research focus of meso-scale urban formed adaptation, the first task is to compare the performance of various urban block types during ESLEs and find out the ones with the strongest capability to reduce the impacts. Although, there are some studies regarding the built potentials and environmental performance of block typologies, no research that focuses on the impact differences on various urban blocks during ESLEs has been found in existing literature. Therefore, this thesis takes it as the first hypothesis.

3.5.2 Hypothesis 2: optimum development density

- There is an optimum development density which might enhance the capacity of waterfronts to deal with ESLEs.

In the previous section, it has been discussed that controlling the development density might reduce the damage caused by ESLEs. However, this issue is complicated by the high market value of the waterfront land. In most cases, urban planners are confronted with the dilemma to balance coastal resiliency and land utilization efficiency. Therefore, the second hypothesis of this thesis is that acertain development density range might balance the waterfront resiliency and land utilization efficiency.

3.5.3 Hypothesis 3: block depth

- It is believed that with other conditions being same, increasing the depth of blocks might reduce the pressure on buildings facing the water.

The third hypothesis is inspired by my recent studies, in which it is found that with other conditions being same, increasing the depth of blocks might reduce the pressure on buildings facing the water. With a quick scan of existing literature, such hypothesis or conclusion is not found. This research considers it as a hypothesis and will check the effect of increasing the depth of blocks on reducing the damage caused by ESLE.

3.5.4 Hypothesis 4: block orientations

- Changing urban blocks' orientation and placing them in an angle to the shoreline can mitigate the damage caused by ESLEs.

The forth hypothesis is developed from Bush et al.'s (1996) study, which has been discussed in a previous section. In their research, Bush et al. (1996) proposed that changing orientations of the waterfront roads and placing them in an angle to the shoreline might help to mitigate the damage caused by ESLEs. However, Bush et al., (1996) admit that: "…road orientation and placement can have a major impact on interior areas flooded, overwashed, and subject to storm surge ebb and they will not, however, reduce frontal wave impacts…" (p.113). In addition, they highlight that these modifications might reduce the traffic flow and suggest that some shore-perpendicular roads can be replaced by mini-parks, playgrounds, and aesthetic plantings.

According to Bush et al.'s (1996) theory, changing urban blocks' orientation will also reduce the impact on buildings so that this study take it as the forth hypothesis. However, this study examines this hypothesis based on not only the interior area flood impact but also frontal wave impacts.

3.5.5 Hypothesis 5: U-shaped blocks

- U-shaped blocks can amplify the impacts caused by ESLEs.

The last hypothesis is developed from Park et al.'s(2013) research, in which they argue that U-shaped blocks can upscale the damage of ESLEs, and it states that separating U-shaped blocks will reduce the impacts during ESLEs.

3.6 Chapter Summary

Unlike other climate change phenomena of climate change, the impacts of sea level rise and ESLEs are significant due to the threat to coastal cities worldwide which can ultimately undermine the entire global economy network if action is not taken. In response to the inevitability of ESLEs, a set of comprehensive adaptation strategies is highly necessary. The adaptation strategies should cover vicarious scales as well as combine top-down and bottom up approaches. More importantly, more attention and effort should be given to transformational adaptation in order to make it more feasible, more acceptable and less costly.

In order to prevent future calamities, urban planners and policy makers should change the way they design and manage the waterfronts. The meso-scaled urban form-based adaptations could play a positive role to respond to ESLEs, while enhancing the resiliency of waterfronts. Since the hypotheses have been raised, in the next chapter, the research methodology employed to examine them will be introduced.

Chapter 4   Research methodology

This chapter describes the research methodology employed in this study. In the first section, the research framework is discussed with the concern of scale and scope of analysis. The second section describes the research design. After elaboration of the research approach, research methods adopted are addressed, including qualitative and quantitative methods, which are followed by the description of main research steps. The third section illustrates the experiment designs used to examining the hypotheses.

4.1   Research scale

For the purpose of clarifying the research methodology, the scale of the analysis in this research is stated first before the discussion of research design and experiment design.

In order to elaborate the research methodology, the foremost issue is to address the scale of analysis. With the purpose of formulating effective adaptations for ES-LEs, research tools and methods for macro-scale (regional, metropolitan level) studies and micro-scale (individual building level) studies have been developed in the past decades. Nonetheless, in the meso-scale realm represented by urban streets and blocks(Ratti & Richens, 2004), well-developed research methodology remains unavailable. Zhang and Heng et al.(2012)argue that together with urban streets and precincts, urban blocks sever as the fundamental components of urban fabric because they are constituted of relatively homogenous buildings, their composition will shape the characteristic of urban public spaces, and they might influence people's perceptions and attitudes to urban environments.Regarding the adaptations to ESLEs, if urban blocks can change the wave behaviors of ES-LEs, they might have significant impacts on the individual buildings and the overall urban fabric in waterfronts. Based on such argument, this research will focus on the meso-scale, in which the influence of urban blocks in waterfronts on wave performance will be studied specifically.

4.2 Research design

In this section, all important issues regarding the research methodology are introduced. The research approach adopted in this study is discussed first, and followed by research methods which include qualitative and quantitative methods. The ways to collect data and summary of research steps are also included in this section.

4.2.1 Research approach

In order to measure the effectiveness of adaptations, this research analyzes the impacts on buildings in different waterfront designs under the same ELSE scenario. If using real cases, the results could not be comparable due to quite various physical contexts between different cases. The surrounding buildings and incongruent spatial settings may have inconsistent impacts on the results so that they may not reflect the exact effectiveness of adaptations. Instead, the research approach combined by typological and parametric analyses is more applicable for this study, in which the theoretical performance of generated urban forms in a simplified physical context can be examined. In other studies with similar research intentions to examine specific built forms' environmental performance, such approach is also widely adopted(Montavon, Steemers, Cheng, & Compagnon, 2006; E. Ng, 2005; Ratti, Raydan, & Steemers, 2003; Zhang et al., 2012).

According to Tice (1993), typology can be defined as the study approach based on objects' shared characteristics which make them part of a larger grouping. Typological research serves as a useful analytical tool as these groupings, or types constitute an essential framework and facilitate comparisons (ibid). In urbanism and architecture studies, typological classification usually is based on physical characteristics such as development intensity, building size, organization pattern, and so on. Having the research purpose to identify effective adaptations in the scope of urban block forms, the primary task is to compare certain basic block typologies and discover which type is best for waterfront development. Besides, other proposed adaptations might also need typological analysis due to their formal characteristics.

A parametric approach is borrowed from statistics, which is a widely-adopted methodology to quantitatively analyze certain variables' influence to the entire system while keeping other settings constant(Geisser & Johnson, 2006). In order to quantify the effectiveness of adaptations, this research will abstract the essentially formal characteristics from each proposed adaptation and parameterize them to some extent while keeping other settings constant. By this method, the capacities of adaptations to interfere with wave activities and mitigate the impact on buildings might be discovered.

Besides, in order to further prove the hypothesis, this research will also carry out an application test. It will be based on the results of the typological and parametric studies. The ultimate target for studies of urban form-based adaptations is to improve the urban design practice in waterfronts and enhance their resiliency. The effective adaptations identified by theoretical experiments might be only partially applicable in the urban design process. Therefore, it is necessary to carry out an application test based on a relatively more complicated and actual urban fabric. The case chosen is the planned waterfront in Singapore for which detailed information will be given in following sections.

4.2.2 Research methods

1) Qualitative methods

In order to achieve the research objectives, the foremost task is to understand the possible impacts of ESLEs on waterfronts developments. Thus, the method of literature review and interview is adopted.

As an important part of the research, reviewing the existing literature regarding waterfront developments is fundamental for this study and must be completed first. With the purpose to argue the importance of this study and to elaborate the research perspective, previous studies about waterfront developments in recent decades are collected and reviewed. In the literature review, the common evolution process of waterfront iselaborated in order to understand the spatial characteristics as well as their possible causes. In addition, the issues of SLR and ESLEs are discussed as serious challenges for waterfront developments, and the actual and possible impacts from SLR and ESLEs on waterfront development are emphasized. Besides, a discussion of various research perspectives for waterfront development in recent decades is delivered so as to elaborate the research perspective of this study.

Moreover, collecting and summarizing the possible adaptations are also key elements of this study, which is focusing on published research. In recent decades, because of the significant consequences of SLR and ESLEs, studies of response strategies, including mitigation and adaptation, attract academic attention in many research areas. With the intention of finding the research gap and formulating the hypothesis, this research provides a large number of achievements and sorts them into various categories.

At last but not least, in order facilitate the application test, interviews with local officials, designers and scholars is also a part of qualitative methods. This research is based in Singapore; therefore, except the summary of published literature regarding the status of ESLEs and waterfront development on this island, the viewpoints from local industry and academia also matter. Local officials in Urban Redevelopment Authority (URA) and Building and Construction Authority (BCA) were asked to give their opinions. In addition, scholars from the School of Design and Environment (SDE), Marina Science Institution (TMSI) and the Center for Sustainable Asian Cities (CSAC) in National University of Singapore (NUS) were also consulted. Some opinions have also been collected during informal discussions with urban designers and architecture students in SDE, NUS.

2) Quantitative methods

Regarding the quantitative method, this study employs simulation tools to simulate the ESLEs and to examine the effectiveness of proposed adaptations.

Simulation of ESLEs

Basically, the simulation of extreme events could be carried out in two methods. On the one hand, it could use physical model tests. For example, Ma et al. (2002) carried out research to study the impacts of 1:1000 year return interval flood of River Calder to the town of Todmorden, West Yorkshire, UK. This method is relatively straightforward in terms of technical requirements. However, it might need quite a large space to settle the physical models and equipment.

On the other hand, the numerical simulation through computers and software is more often adopted in this kind of studies. It requires numerical models and software which could simulate the events' outcome. In the process of simulation, the parameters of interest, such as velocity, pressure, wave height and energy can be calculated, recorded and illustrated either through diagrams, charts or even through animations. Obviously, this method depends more on the quality of the computers and software. Nowadays, due to the rapid development of computer technology, the performance of numerical simulation has improved significantly. Some complicated simulations which researchers could not imagine before now became feasible. Although the simulation software of flow dynamics still has a lot of outstanding challenges, its advantages are quite evident. First of all, it is easier to change the initial settings as it allows a broader application. In addition, the results from numerical simulation are more visually intuitive since they are recorded automatically in several formats which make the comparison simpler.

Considering the above, this study will employ numerical simulation as the main research tool. The numerical simulation is an important method to understand the activity pattern and potential impact of catastrophic events so that it becomes the focus of researchers in many disciplines. For example, Caleffi et al (2003) simulated the potential flood of Toce Rive through two dimensional shallow water equations. Haider et al(2003) used the same equations and ran a numerical simulation on the city of Nimes in France. Yim(2005) and his colleagues in Oregon State University discussed tsunami and storm surge hydrodynamic loads on coastal bridge structures through numerical model simulation.

Computational fluid dynamics (CFD)

Numerical simulation contains fluid dynamics, an important branch which is called-computational fluid dynamics (CFD) and is effective to solve and analyze problems that involve fluid flows. Computers are used to perform the calculations required to simulate the interaction of liquids and gases with surfaces defined by boundary conditions(J. D. Anderson & Wendt, 1995). Currently, researchers have developed many variations of CFD software and some of them are commercialized and widely adopted in the studies of engineering design, chemical engineering, aircraft and automobile manufacturing, ventilation system design, and marina engineering. However, considering that most CFD codes have been developed for mechanical and chemical engineering, the application of CFD to hydraulic problems raises a number of issues related to geometry, boundary conditions, hydraulic roughness, flow regime and turbulence (Pender, Morvan, Wright, & Ervine, 2005). Besides, ESLEs simulation is generally called free surface flow, which is difficult as a free-boundary problem. A free boundary poses the difficulty that on the one hand the solution region changes when its surface moves, and on the other hand, the motion of the surface is in turn determined by the solution (Monaghan, 1994). Considering that it is designed for architecture and urban study and its strengths in free surface flow, this research chooses scSTREAM which is developed by Cradle as the simulation tool.

Using scSTREAM to simulate ESLEs

Similar to the other CFD software, the basic consideration of scSTREAM is to deal with the continuous fluid in a discrete way by using computers, and has three components, including preprocessor, solver, and postprocessor. Its processing could also be divided into three major steps, namely preprocessing, processing, and post-processing. During preprocessing, except for building the 3-dimensional model, the geometry domain of the problem, and the boundary conditions (sometimes together with initial conditions), the physical modeling such as the equation of turbulent, heat, radiation, and so on are also defined. Meanwhile, the volume occupied by fluid is going to be divided into discrete cells, which is normally called meshing. In addition, scSTREAM allows users to set certain variables to be recorded during processing, such as velocity, pressure, and temperature. The solver may start the processing when these settings are done. The specific equations are iteratively solved by computers. After solving them, the postprocessor can be used as a tool to analyze and visualize results. The process of simulations is summarized in Figure 4.1.

Figure 4.1 shows the inputs and outputs during the entire process. For preprocessing, files generated by common design software, such as AutoCAD, Sketchup, and 3DMax can be directly or indirectly imported into Preprocessor for the purpose of building 3 dimensional models. The preprocessor can generate CAB files which contain all information about 3D models and other settings for simulations. Meanwhile, it also generates S files which can be imported into solver and the solver can start simulations based on it. Normally, three kinds of files can be outputted by solver, namely L file, FLD file and TM file. The L file is an important output which contains the data for further analysis, in which variables (velocity, wave height, pressure, temperature, and so on) are recorded and listed. This research uses L file to analyze the pressure on buildings caused by ESLEs. The FLD file is also a record of important variables and it is intended to be inputted into postprocessor for visualization. The TM file saves instantaneous values of velocity, pressure, and temperature atarbitrary specified points. With the input of a series of FLD files, the postprocessor can generate videos and images. Meanwhile, it can also generate a Status file, which is used for saving the settings for visualization status, including the information of perspective, render options, light, and more.

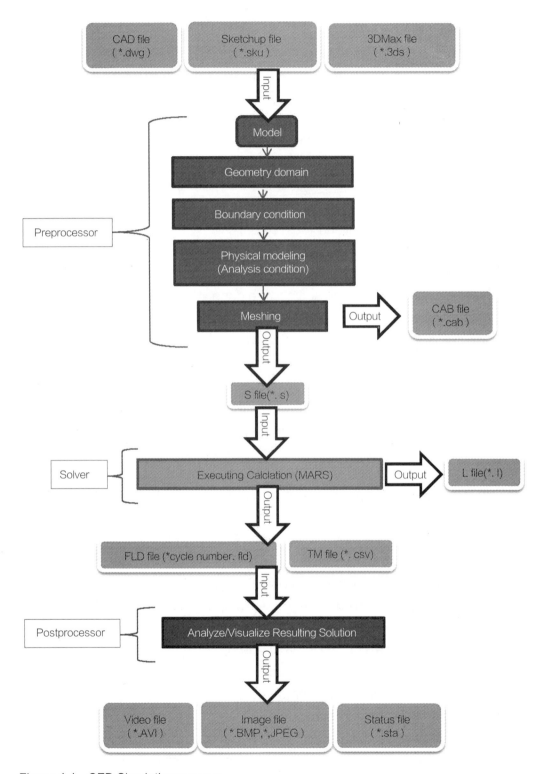

Figure 4.1   CFD Simulation process

The ESLEs simulations in this research use the function of "Free Surface" in scSTREAM, which is relatively more complicated because it contains two kinds of fluids, namely air and water. The Multi-interface Advection and Reconstruction Solver (MARS) methods are adopted, which is one of the latest and sophisticated methods capable of capturing free surface geometries with high accuracy. The MARS method can analyze gas-liquid or liquid-liquid types of two-phase flows. At the same time, the phenomena with bubbles rising in the liquid tank and droplets splitting or colliding in the opposite phase can be considered. Between two types of analyses, steady-stat analysis and transient analysis, this research chooses the latter one, which means the simulations do not capture single moments but lasts for a short period, ranging from 60 to 200 seconds.

4.2.3 Data collection

The data required for simulations is mainly summarized from literature. In regard to the application test based on Singapore's scenario, interview and field work is also necessary. Firstly, governmental agencies have released some information for future waterfront development. With them, the digital model of waterfront development proposals can be built up for simulations. Secondly, the studies of future sea level rise and ESLEs in Singapore have already been completed recently, and the conclusion is similar to the publications of Intergovernmental Panel in Climate Change (IPCC). With the help of faculty members in Department of Civil and Environmental Engineering in NUS and scholars in TMSI, necessary data about SLR and ESLEs was acquired.

During simulations, the data for analysis can be recorded automatically. In this study, the measurement of the effectiveness of adaptations is based on the recorded pressure on the facades. Therefore, as shown in Figure 4.2, for some façades facing water wave the pressure is registered as "PartFaces" by the function of "Edit Part Face" during preprocessing, which enables the solver to record the force caused by ESELs on façades. In preprocessor, the last cycle number is defined for every simulation, which means to set how long the simulation may last. The period of a simulation is defined to make sure that the waves can reach all the buildings and last for a while. Therefore, for the simulations in this study, there are at least 5000 and at most 8000 cycles. For every cycle, the force values on the registered facades is recorded in the generated L files on three axes (Force-X, Force-Y, and Force-Z), as shown in Figure 4.2.

More details regarding simulation data analysis are introduced in the next section.

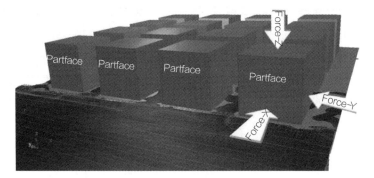

Figure 4.2   Registered Partfaces to record the forces Force-x, Force-y, and Force-z

4.2.4   Research steps and research framework

To sum up, this research takes six steps to verify the hypothesis:

• To summarize possible meso-scale urban block-based adaptations as through literature review and interviews.

• To simplify these adaptations into typological models based on their main physical characteristics.

• To create a series of modelsintegrated with adaptations to a variable extent.

• To simulate ELSEs on these typological models and compare the impacts on building facades.

• To identify effective adaptations through comparisons.

• To apply the identified adaptations on an urban design proposal of waterfront development in Singapore in order to further test their effectiveness and feasibility.

The research framework is shown in Figure 4.3.

4.3   Experiment design

As it is shown in Figure 4.4, there are four steps to carry out the experiments. The

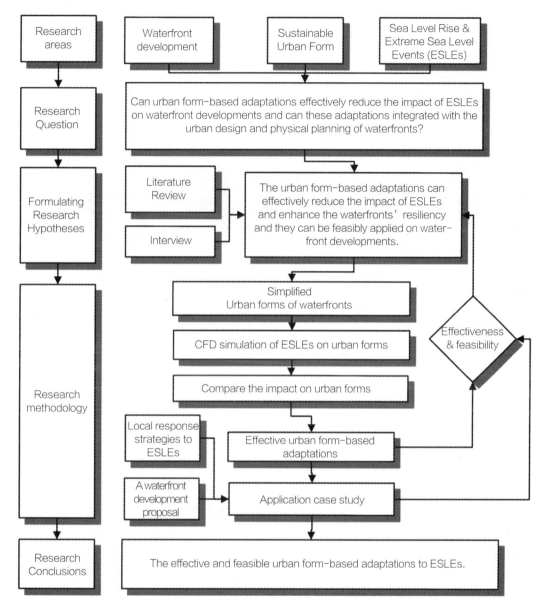

Figure 4.3   Research framework

first one is to create the simplified models of urban waterfront with a high sea level and a sea wall. Secondly, an ESLE is simulated on the models, which is considered as a severe storm surge flood. The third step is to enable the ESLE breaks through the sea wall and their impact on buildings. Finally, some façades facing the waves are registered in order to record the pressure on buildings for further analyses.

In order to examine the hypotheses, there are a total of five experiments designed, namely experiment for block typologies, experiment for density controlling, experiment for block length, experiment for building orientations, and experiment for U-shaped blocks. The details of these experiments are illustrated later in this section after the introductions of the variables and criterion for comparisons and the setting of urban context and ESLEs' scenario.

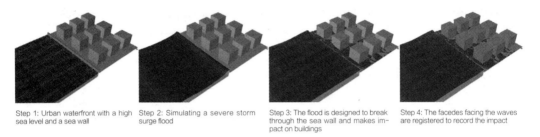

Step 1: Urban waterfront with a high sea level and a sea wall

Step 2: Simulating a severe storm surge flood

Step 3: The flood is designed to break through the sea wall and makes impact on buildings

Step 4: The facades facing the waves are registered to record the impact

Figure 4.4　Four steps to carry out experiments

4.3.1　Variables and criterion

The variables to examine adaptations' effectiveness are maximum Force-X and Force-X per unit width (FXPUW). When the waves of ESLEs reach the buildings in waterfronts, they generate massive force on the façades facing them. The function of adaptations is to reduce the force and mitigate the damage. Therefore this study measures the adaptations' effectiveness through the comparison of forces on facades between models. As described in last section and shown in Figure 4.2, the force is recorded on three axes; however, this study only takes the Force-x (on Axisx) as a main consideration since it is much stronger than Force-y and ESLEs cannot generate any impact in Axis-z. For some experiments and the application test, the areas of facades might be different so that the Force-x is not comparable. Thus, considering that the building heights in experiments may vary and the waves only have impact on the lower part of buildings, maximum FXPUW is also identified to be the main variable to measure adaptations' effectiveness in certain experiments. The maximum FXPUW is the ratio of maximum Force-x to the projected width of buildings on y axis.

During simulations, up to 5,000-8,000 values of Force-x are going to be recorded. It is impossible to compare the values of every cycle. Therefore, the research chooses maximum values and average values of Force-x or FXPUW to be the variables. Between these two, maximum values are identified as the primary variable because it is believed to generate the most significant damage to buildings.

The criterion for comparison is quite straightforward: if the force generated by ES-LEs decreases, it is believed that the proposed adaptation is effective. In a certain experiment, the maximum values of Force-X or FXPUW are selected first for comparisons. The designed model with the lowest value of maximum Force-X or FXPUW is identified as the best solution. If this model was designed with an adaptation, it means this adaptation is effective and vice versa. Secondly, average values are also compared, but only for reference.

4.3.2 The setting of urban context and ESLEs' scenario

The urban context is designed as an urban waterfront with a high sea level. The sea level is designed as 1m lower than the waterfront land elevation. Meanwhile, a seawall is designed as 2m higher than sea level and 1m higher than the waterfront land elevation(Figure 4.5).

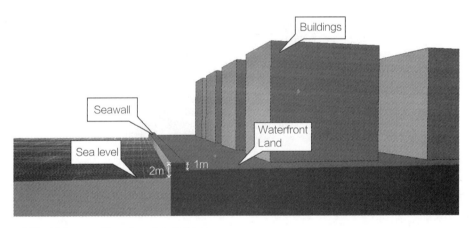

Figure 4.5   Urban setting for simulations

The ESLEs' scenario in experiments is generated by the function of "Wave Generation Source (WGS)", which is designed as a severe storm surge flood with wave height=5 meters and water period=10 seconds. The detailed setting is shown in the Figure 4.6. When the waves generated by WGS method reach the boundary, they do not completely flow out of the boundary; some of them might reflect on the boundary and travel backward, which will cause appearance of unnecessary waves within the analysis region and deteriorate simulation accuracy. Therefore, advised by the developer of scSTREAM, an "Attenuation Zone" is also created behind the WGS in order toprevent unnecessary waves. The position of Attenuation Zone is shown in Figure 4.6 and its setting is shown in Figure 4.7.

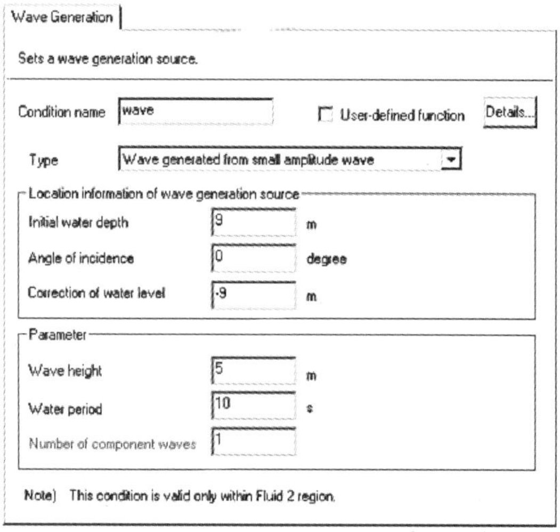

Figure 4.6 Settings for wave generation source (WGS)

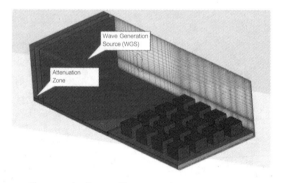

Figure 4.7 Wave generation and attenuation zone

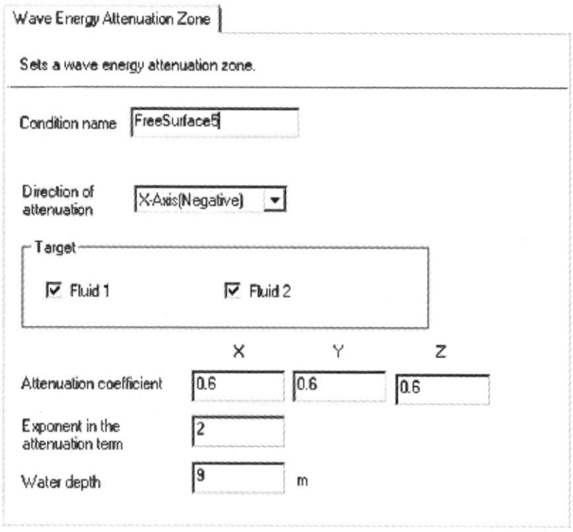

Figure 4.8 Setting for attenuation zone

4.3.3 Experiment 1 for block typologies

The purpose of this experiment is to examine the first hypothesis, to compare various urban block typologies, and to and identify one certain typology with best capacity to mitigate the impact from ESLEs. Actually, in 1960s, the study of urban block typology and archetype was initiated by Martin and March (1972), which focuses on the question of "What building forms make the best use of land?" (Ratti et al., 2003, p. 49), which attracted the attention of planners and architects. Since then, their research inspires more scholars(Blowers, 2013; VK Gupta, 1984; Vinod Gupta, 1987; Steemers et al., 1997) and quite a lot of archetypes have been developed to address the characteristics of urban fabric and texture in order to enhance the environmental behavior and facilitate the process of urban design. Initially, such studies focused on only two forms, namely courtyards and pavilions(Figure 4.9), with the intention to create more open space, enhance sunshine lighting, and reduce the height of building(Ratti et al., 2003). Subsequently, in addition to the pavilion and courtyard, March and Trace (1968) take another block typology into account by considering the elementary form of street, which is the slab. Through combinations of these basic block typologies, they generated six archetypal forms (Figure 4.10).

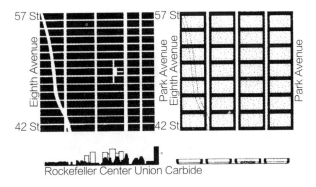

Figure 4.9 Two basic block forms——pavilion (left) and courtyard (right), the case study of Manhattan
Source: Ratti, et al., 2003, p. 50.

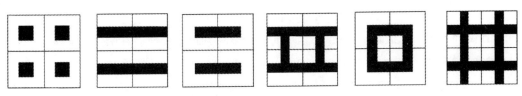

Figure 4.10 Plans for six typological forms of urban blocks created by March and Trace
Source: Ratti, et al., 2003, p.51.

In the studies introduced above, the research objectives are to examine the built potentials, such as site coverage and building height, and the environmental behaviors(energy efficiency, daylight availability and ventilation capacity) of these block typologies. The typologies designed in this research are based on typological forms created by March and Trace (Figure 4.9), whereas, this research focuses on their capacities to mitigate the impact of ESLEs so that the inner space of urban blocks might not be considered. Therefore, the typology of courtyard is not included. As shown in Figure 4.11, seven typologies will be analyzed in this experiment, including small pavilions (Model SP), big pavilions (Model BP), mega pavilions (Model MP), transverse slabs (Model TS), short transverse slabs (Model STS), longitudinal slabs (Model LS),andshort longitudinal slabs (Model SLS).The quantity of gross floor area in all models equates.

The simulations are designed to reach 6000 cycles and last for around 120 seconds. The detail experiment setting is given in appendix A. All facades facing the waves are registered as PartFaces to record Force-X. Because the building units are same in terms of shape and scale, maximum Force-X is selected for comparison.

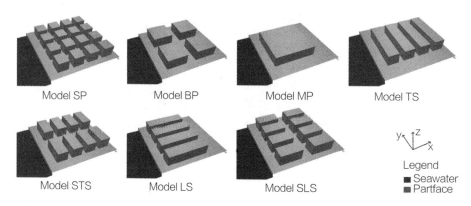

Figure 4.11 All models in Experiment 1

4.3.4 Experiment 2 for optimum development density

For the purpose of waterfront development practices, the main objective of this experiment is to examine the second hypothesis and develop a method for urban planners to formulate the proper site coverage in vulnerable coastal areas. In other words, Experiment 2 is intended to reveal the optimum site coverage value for waterfront developments, in which simplified grid urban layouts with typical values of site coverage are designed for comparison.

Site coverage is chosen to be the parameter that defines development densities. Site coverage, also known as building coverage ratio, coverage ratio, building coverage, building footprint, and lot coverage, refers to the ratio between the ground floor area of the building or buildings and the site area. It is an essential parameter to control urban density and development intensity.

The construction lot is designed as a rectangle with a 1000m²(125m × 80m) area. In order to facilitate the comparison and analysis, the building unit is identically designed as a cube (20 × 20 × 20m³), which might represent the scale of individual houses. Through different combinations (Model A, B, C, D, E and F) of the building units, the values of site coverage in the same lot are respectively 16%, 24%, 36%, 48%, 60%, and 80% as shown in the Figure 4.12. These samples covered most of the typical site coverage values in real urban developments.

Considering the scale for this experiment is relatively small, the simulations are designed with 5000 cycles and approximately 150 seconds. The detail experiment setting is given in appendix B. All facades facing the waves are registered as Part-Faces to record Force-X. Because the building units are same in terms of shape and scale, maximum Force-X is selected for comparison.

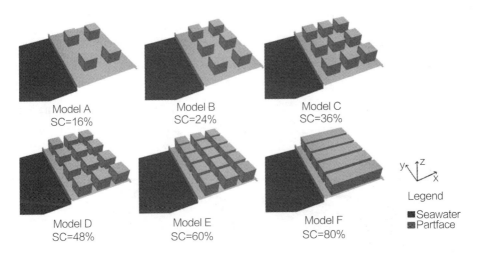

Figure 4.12   All models in Experiment 2

4.3.5 Experiment 3 for block depths

Experiment 3 is inspired by the results of Experiment 1 and 2. In the first two experiments, it is found that with other conditions being same, increasing the depth of blocks might reduce the pressure on buildings facing the water. With a quick scan of existing literature, such hypothesis or conclusion is not found. The research considers it as the third hypothesis and designs the Experiment 3 to examine it. In other words, the objective of Experiment 3 is to check the effect of increasing the depth of buildings.

If the simplified buildings in Experiment 2 represent individual houses, the building scale in this and subsequent experiments increases and represents commercial, public, and congregated residential buildings. In order to examine the impact of increasing the block length, the building units are designed as cube with 35 m width, which refers to the common block scale in Singapore' downtown area. Since this research is based on Singapore, a sketched analysis of block scales in CBD area was carried out before this research and it revealed that 35m is a common width for most of urban blocks. In addition, it also found that the sizes of 60m and 135m are also quite frequently chosen for the block depth in CBD area. Based on this statistic, the width for all blocks is designed as 35m, while three values are chosen for the depth of the blocks, namely 25m, 60m, and 135m. Because the height of the building might not influence the result if it will be higher than waves, it is designed as 50m for all units. As the building units' depth reduces, the quantity of units will be increased in order to ensure density does not vary too much in different models and affect the results.

As shown in Figure 4.13, three models are designed in this experiment, namely Model G, Model H, and Model I. The model G is composed of 32 smaller units (width=35m, depth=25m, height=50m), while Model H is composed of 16 relatively longer units (width=35m, depth=60m, height=50m), and the buildings in Model I have the longest depth (width=35m, depth=135m, height=50m).

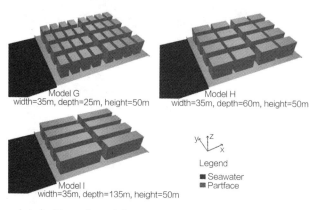

Figure 4.13   All models in Experiment 3

The simulations are designed with 8000 cycles and last for less than 200 seconds. The detailed experiment setting is given in appendix C. Only the facades with the same locations are registered as PartFaces to record Force-X (Figure 4.13). Because the area of all registered PartFaces equate with each other, maximum Force-x is selected for comparison.

4.3.6 Experiment 4 for block orientations

As discussed in the previous chapter, one hypothesis in this thesis is that changing block orientations might also help to mitigate the damage caused by ESLEs. Therefore, Experiment 4 is designed to examine the effect of changing the angle between the long side of buildings and the shoreline, which is named as α in order to facilitate the narrative of this experiment as shown in Figure 4.14.

Experiment 4 is more complicated than others, in which 4 models are created as shown in the Figure 4.14, including Model J, K, L, and M. Model J is modified based on the Model H in Experiment 3, which has nine buildings with their long side perpendicular to the shoreline, which means $\alpha = 0°$. In Model K, there are also nine buildings but the angle α increased to 15° whilethe angle α increases to 30° in Model L and to 45° in Model M. With the orientation of the urban block being rotated, some building's shapes also change. Therefore, in order to reduce the influence caused by increasing the depth of buildings, the layouts of urban blocks are restructured in Model L and M. As shown in Figure 4.14, if the depth is too long for a single building, it will be split into two; meanwhile, if the depth is too short for a building, it will be merged into another building and form a relatively bigger unit.

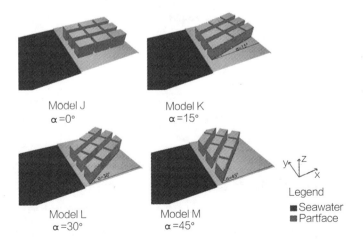

Figure 4.14   All models in Experiment 4

Since there are two façades of each building that may bear the Force-X caused by waves in model K, L and M, and the area of these facades is not equal, the maximum FXPUW is considered as the variable to compare the impacts from ESLEs. The simulations are designed with 8000 cycles and last for less than 160 seconds. The detail experiment setting is given in appendix D. As shown in Figure 4.14, only the facades with relatively same locations are registered as PartFaces to record Force-x.

4.3.7 Experiment 5 for U-shaped blocks

The last hypothesis which will be examined in this research is regarding different types of U-shaped blocks. The U-shaped block is quite often constructed in waterfronts, especially for leisure and entertainment functions, such as hotels and restaurants, because it can provide more waterscapes for users. However, as described in the last chapter, this kind of blocks might have a negative effect during the ESLEs and amplify the impacts from waves. Therefore, this experiment is designed to examine several U-shaped blocks.

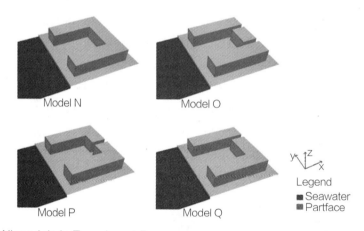

Figure 4.15   All models in Experiment 5

As it can be seen in Figure 4.15, four types of U-shaped blocks are created, namely Model N, O, P, and Q. Among them, Model N is a normal U-shaped Block in which all buildings are connected with each other; while the design for other models is intended to break the connection to certain extent. In Model O, all three parts of the U-shaped block are separated, which make this block a combination of two larger buildings on the sides and one smaller building in the middle. In Model P, the block is cut in the middle and becomes two completely symmetricalparts. In Model Q, one leg of the U-shaped block is separated from the other parts, which forms two parts, one rectangle building and one L-shaped building.

The simulations are designed with 7000 cycles and last for less than 200 seconds. The detail experiment setting is given in appendix E. All facades facing the waves are registered as PartFaces to record Force-x. As shown in Figure 4.15, the facades' areas on the two legs of all models are equal. But regarding the middle part of U-shaped block, the areas of facades are not same, so the FXPUW is considered as the variable to compare the impacts of ESLEs in this experiment.

4.4 Application study

In order to examine the feasibility of proposed adaptations, an application test is necessary to be carried out. After the proposed hypotheses being examined, some of them might be proved theoretically effective. However, because the models in all experiments are created based on the physical characteristics extracted from the hypotheses, to apply these adaptations into actual urban design is still an important matter. Therefore, this study will conduct an application case study in order to integrate the meso-scale urban form-based adaptations with waterfront design projects.

The methodology to carry out the application study is quite similar with the experiments. Because the scale of the model is increased significantly, some specific settings are changed for simulations. The case is chosen in the city of Singapore and based on the design proposal of Marina South that is a waterfront development located in urban center. The detailed information regarding the application study will be presented in Chapter 6.

4.5 Chapter summary

In this chapter, the research methodology used to examine the proposed hypotheses and 5 experiment designs have been elaborated. In the next chapter, the results of all experiments are illustrated and analyzed. With the analysis of results, the hypotheses will be proved.

Chapter 5 Results and hypotheses examination

This chapter reports the results of five experiments through charts and tables. In the previous chapter, in order to examine five proposed hypotheses, five experiments were designed. Through a large number of simulations, the research gained a great number of test results. In this chapter these results are summarized and analyzed, and the hypotheses will be verified accordingly. After analyzing the experiment results of the five experiments in sequence, the discussion in regard with the applications of meso-scale urban form based adaptation will be presented at the end of this chapter.

5.1 Results of Experiment 1

The purpose of this experiment is to examine the first hypothesis and compare various urban block typologies and identify the one with best capacity to mitigate the impact from ESLEs. In this experiment seven typologies (models) will be analyzed, including small pavilions (Model SP), big pavilions (Model BP), mega pavilions or courtyards (Model MP), transverse slabs (Model TS), short transverse slabs (Model STS), longitudinal slabs (Model LS), and short longitudinal slabs (Model SLS), as introduced in section 4.3.3. The quantity of gross floor area in all models equates, and the simulations are designed to reach 6000 cycles and last for around 120 seconds. All facades facing the waves are registered as PartFaces to record Force-x. Because all PartFaces are the same in terms of shape and scale, maximum Force-x is selected for comparison.

5.1.1 Model SP

As shown in Figure 5.1, a total of 16 facades facing waves in Model SP are registered as PartFaces to record Force-x. During 120-seconds simulation, 6000 values of Force-x are recorded as shown in appendix G. The maximum and average values of Force-x on all facades are summarized in Table 5.1. The maximum Force-x on every facade will be compared with other models in this experiment, while the average Force-x is for reference only.

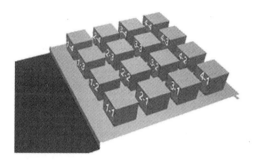

Figure 5.1 Sixteen facades are registered as PartFaces in model SP

Maximum and average Force-x on all facades in Model SP   Table 5.1

	f 1-1	f 1-2	f 1-3	f 1-4	f 2-1	f 2-2	f 2-3	f 2-4
Maximum Force-x (N)	1.64 E+06	1.55 E+06	1.94 E+06	2.39 E+06	8.49 E+05	8.53 E+05	9.62 E+05	1.05 E+06
Average Force-x (N)	7.07 E+05	7.11 E+05	7.47 E+05	7.61 E+05	5.38 E+05	5.41 E+05	5.50 E+05	5.58 E+05
	f 3-1	f 3-2	f 3-3	f 3-4	f 4-1	f 4-2	f 4-3	f 4-4
Maximum Force-x (N)	6.98 E+05	7.17 E+05	7.67 E+05	8.30 E+05	5.44 E+05	5.42 E+05	5.54 E+05	5.69 E+05
Average Force-x (N)	4.88 E+05	4.89 E+05	4.93 E+05	5.04 E+05	4.22 E+05	4.23 E+05	4.30 E+05	4.41 E+05

5.1.2 Model BP

As shown in Figure 5.2, a total of eight facades facing waves in Model BP registered as PartFacesto record Force-x. Similarly, 6000 values of Force-x are recorded as shown in Appendix G. The maximum and average values of Force-x on all facades are summarized in Table 5.2. The maximum Force-x on every facade will be compared with other models in this experiment, while the average Force-x is for reference only.

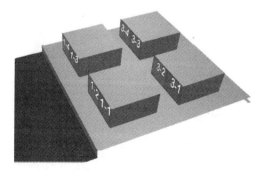

Figure 5.2 Eight facades are registered as PartFaces in model BP

Maximum and average Force-x on all facadesin Model BP　　　　　　　　　　　Table 5.2

	f 1-1	f 1-2	f 1-3	f 1-4	f 3-1	f 3-2	f 3-3	f 3-4
Maximum Force-x (N)	1.73 E+06	1.48 E+06	2.03 E+06	2.49 E+06	7.84 E+05	7.87 E+05	8.09 E+05	8.42 E+05
Average Force-x (N)	6.89 E+05	6.71 E+05	7.37 E+05	7.68 E+05	4.94 E+05	4.95 E+05	5.01 E+05	5.01 E+05

5.1.3　Model MP

As shown in Figure 5.3, a total of four facades facing waves in Model MP registered as PartFaces to record Force-x. The simulation lasts for 120 seconds, and 6000 values of Force-x are recorded as shown in Appendix G. The maximum and average values of Force-x on all facades are summarized in Table 5.3. The maximum Force-x on every facade will be compared with other models in this experiment, while the average Force-x is for reference only.

Figure 5.3　Four facades are registered as PartFaces in model MP

Maximum and average Force-x on all facades in Model MP　　　　　　　　　　Table 5.3

	f 1-1	f 1-2	f 1-3	f 1-4
Maximum Force-x	1.51E+06	1.46E+06	1.81E+06	2.25E+06
Average Force-x	6.73E+05	6.89E+05	6.92E+05	6.86E+05

5.1.4　Model TS

As shown in Figure 5.4, a total of 16 facades facing waves in Model TS registered as PartFaces to record Force-x. The simulation lasts for 120 seconds and 6000 values of Force-x are recorded as shown in Appendix G. The maximum and average values of Force-x on all facades are summarized in Table 5.4. The maximum Force-x on every facade will be compared with other models in this experiment, while the average Force-x is for reference only.

Figure 5.4 Sixteen facades are registered as PartFaces in model TS

Maximum and average Force-x on all facades in Model TS   Table 5.4

	f 1-1	f 1-2	f 1-3	f 1-4	f 2-1	f 2-2	f 2-3	f 2-4
Maximum Force-x (N)	1.48 E+06	1.45 E+06	1.80 E+06	2.20 E+06	6.92 E+05	6.50 E+05	6.88 E+05	6.22 E+05
Average Force-x (N)	6.86 E+05	7.06 E+05	7.31 E+05	7.37 E+05	4.58 E+05	4.56 E+05	4.51 E+05	4.48 E+05
	f 3-1	f 3-2	f 3-3	f 3-4	f 4-1	f 4-2	f 4-3	f 4-4
Maximum Force-x (N)	6.60 E+05	6.65 E+05	6.35 E+05	6.64 E+05	4.89 E+05	4.76 E+05	4.66 E+05	4.89 E+05
Average Force-x (N)	4.56 E+05	4.62 E+05	4.59 E+05	4.52 E+05	4.04 E+05	4.05 E+05	4.01 E+05	4.12 E+05

5.1.5 Model STS

As shown in Figure 5.5, a total of 16 facades facing waves in Model BP registered as PartFaces to record Force-x. The simulation also lasts for 120 seconds, 6000 values of Force-x are recorded as shown in Appendix G. The maximum and average values of Force-x on all facades are summarized in Table 5.5. The maximum Force-x on every facade will be compared with other models in this experiment, while the average Force-x is for reference only.

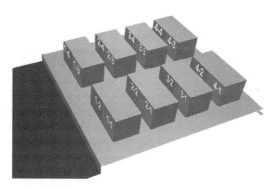

Figure 5.5 Sixteen facades are registered as PartFaces in model MP

Maximum and average Force-x on all facades in Model STS Table 5.5

	f 1-1	f 1-2	f 1-3	f 1-4	f 2-1	f 2-2	f 2-3	f 2-4
Maximum Force-x (N)	1.70 E+06	1.48 E+06	1.98 E+06	2.44 E+06	8.32 E+05	8.92 E+05	7.97 E+05	9.06 E+05
Average Force-x(N)	7.18 E+05	6.94 E+05	7.37 E+05	7.46 E+05	5.24 E+05	5.22 E+05	5.21 E+05	5.20 E+05
	f 3-1	f 3-2	f 3-3	f 3-4	f 4-1	f 4-2	f 4-3	f 4-4
Maximum Force-x(N)	7.02 E+05	7.05 E+05	7.17 E+05	7.45 E+05	5.40 E+05	5.33 E+05	5.63 E+05	5.57 E+05
Average Force-x(N)	4.84 E+05	4.85 E+05	4.95 E+05	4.96 E+05	4.20 E+05	4.19 E+05	4.35 E+05	4.34 E+05

5.1.6 Model LS

As shown in Figure 5.6, a total of four facades facing waves in Model BP registered as PartFaces to record Force-x. The simulation lasts for 120 seconds and 6000 values of Force-x are recorded as shown in Appendix G. The maximum and average values of Force-x on all facades are summarized in Table 5.6. The maximum Force-x on every facade will be compared with other models in this experiment, while the average Force-x is for reference only.

Figure 5.6 Four facades are registered as PartFaces in model LS

Maximum and average Force-x on all facades in Model LS Table 5.6

	f 1-1	f 1-2	f 1-3	f 1-4
Maximum Force-x (N)	1.42E+06	1.27E+06	1.90E+06	2.34E+06
Average Force-x (N)	6.40E+05	6.51E+05	6.94E+05	7.25E+05

5.1.7 Model SLS

As shown in Figure 5.7, a total of four facades facing the water in Model BP registered as PartFaces to record Force-x. The simulation lasts for 120 seconds and 6000 values of Force-x are recorded as shown in Appendix G. The maximum and average values of Force-x on all facades are summarized in Table 5.7. The maximum Force-x on every facade will be compared with other models in this experiment, while the average Force-x is for reference only.

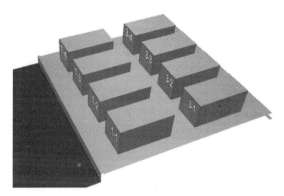

Figure 5.7   Eight facades are registered as PartFaces in model SLS

Maximum and average Force-x on all facades in Model SLS   Table 5.7

	f 1–1	f 1–2	f 1–3	f 1–4	f 3–1	f 3–2	f 3–3	f 3–4
Maximum Force-x (N)	1.62 E+06	1.52 E+06	1.91 E+06	2.36 E+06	8.80 E+05	7.75 E+05	8.96 E+05	9.70 E+05
Average Force-x (N)	6.94 E+05	6.96 E+05	7.24 E+05	7.66 E+05	5.07 E+05	5.07 E+05	5.09 E+05	5.16 E+05

5.1.8 Comparison and analysis

In this experiment, under the same extreme sea level event scenario, a total of seven models are tested. As shown in Table 5.8, maximum values of all facades facing the water waves in all models are summarized. Due to different combinations, the quantity of facades varies between these models. For instance, there are 16 facades in the Model SP, TS and STS, while only four facades in the Model MP and LS; and other models may have eight facades. It has been noticed in all simulations that from buildings in the front to the buildings in the back, the values of maximum Force-x reduce significantly. Therefore, the comparison will be carried out according to the relative positions of facades.

Maximum Force-x (N) of all models in Experiment 1   Table 5.8

	f 1-1	f 1-2	f 1-3	f 1-4
Model SP	1.64E+06	1.55E+06	1.94E+06	2.39E+06
Model BP	1.73E+06	1.48E+06	2.03E+06	2.49E+06
Model MP	1.51E+06	1.46E+06	1.81E+06	2.25E+06
Model TS	1.48E+06	1.45E+06	1.80E+06	2.20E+06
Model STS	1.70E+06	1.48E+06	1.98E+06	2.44E+06
Model LS	1.42E+06	1.27E+06	1.90E+06	2.34E+06
Model SLS	1.62E+06	1.52E+06	1.91E+06	2.36E+06
	f 2-1	f 2-2	f 2-3	f 2-4
Model SP	8.49E+05	8.53E+05	9.62E+05	1.05E+06
Model BP	N/A	N/A	N/A	N/A
Model MP	N/A	N/A	N/A	N/A
Model TS	6.92E+05	6.50E+05	6.88E+05	6.22E+05
Model STS	8.32E+05	8.92E+05	7.97E+05	9.06E+05
Model LS	N/A	N/A	N/A	N/A
Model SLS	N/A	N/A	N/A	N/A
	f 3-1	f 3-2	f 3-3	f 3-4
Model SP	6.98E+05	7.17E+05	7.67E+05	8.30E+05
Model BP	7.84E+05	7.87E+05	8.09E+05	8.42E+05
Model MP	N/A	N/A	N/A	N/A
Model TS	6.60E+05	6.65E+05	6.35E+05	6.64E+05
Model STS	7.02E+05	7.05E+05	7.17E+05	7.45E+05
Model LS	N/A	N/A	N/A	N/A
Model SLS	8.80E+05	7.75E+05	8.96E+05	9.70E+05
	f 4-1	f 4-2	f 4-3	f 4-4
Model SP	5.44E+05	5.42E+05	5.54E+05	5.69E+05
Model BP	N/A	N/A	N/A	N/A
Model MP	N/A	N/A	N/A	N/A
Model TS	4.89E+05	4.76E+05	4.66E+05	4.89E+05
Model STS	5.40E+05	5.33E+05	5.63E+05	5.57E+05
Model LS	N/A	N/A	N/A	N/A
Model SLS	N/A	N/A	N/A	N/A

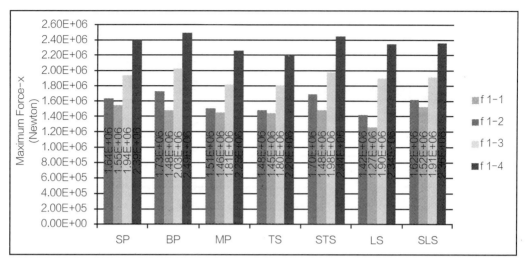
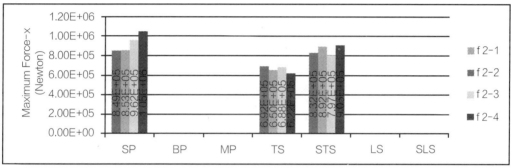
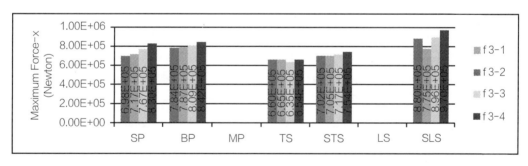
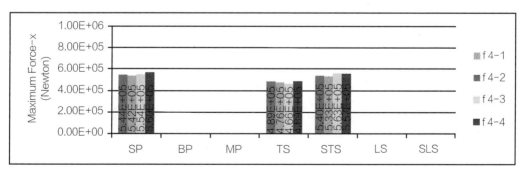

Chart 5.1 Comparison of Force-x on facades

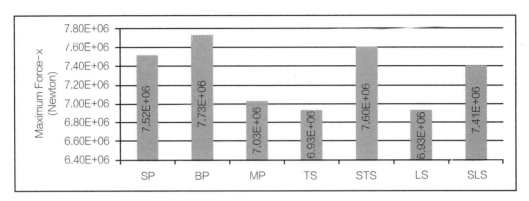

Chart 5.2 Comparison of sum of maximum Force-x in the first row

Chart 5.1 shows the comparison of maximum Force-x among seven models. Compared with other rows, in the first row, not only the values are much higher but also the difference between models is more significant, which means that the buildings in the front of waterfronts will bear the brunt of ESLEs. Therefore, it is believed that the foremost purpose to applying adaptations is to reduce the impact on the buildings in the front as much as possible. Therefore, this research will emphasize the comparison of the impact on the buildings in the front.

For the fourPartFaces (f1-1, f1-2, f1-3, and f1-4) in the first row, the maximum values vary widely. Among them, the values on façade f1-4 are higher than on other facades in all models. Meanwhile, values on f1-2 are below other facades. As shown in Chart 5.1, the trend is generally noticeable between these models. For f1-1, the highest value of maximum Force-x (1.73E+06 N) is found in model BP; the lowest value (1.42E+06 N) is found in Model LS. Regarding the impacts on f1-2, the highest value is found in Model SP (1.55E+06 N); while the lowest value (1.27E+06 N) is also found in Model LS. For f1-3, the highest value (2.03E+06 N) is found in Model BP, and the lowest value (1.80E+06 N) is found in Model TS. On f1-4, the highest value (2.49E+06) also exists on Model BP and lowest Force-x is found in Model TS (2.20E+06 N).

Although these maximum forces may not appear simultaneously, in order to evaluate the total performance of all models, the values of maximum Force-x on all facades in the first row are combined together and compared, as shown in Chart 5.2. Overall, in terms of buildings at the very front, the performances of Model TS and LS are better than others; the combined values of maximum Force-x of both models are 6.93E+06 N. On the order hand, the performance of Model BP is the worst among all models, whose sum of maximum Force-x is 7.73E+06. The total force generated by waves can be reduced by 8.00E+05 N (10.3%) if urban blocks change their topology from BP to TS or LS. Between Model TS and LS, the latter one may be better than former one because Model LS only has four facades facing waves but Model TS has 16 facades.

In the second and forth row, only three models (SP, TS and STS) can be compared, while in the third row, five models (Model SP, BP, TS, STS, and SLS) can be compared. Compared to the first row, the values and differences are not significant as ones in the first row. As shown in chart 5.1, it is quite clear that the maximum values for all facades on Model TS are lower than others, which means the performance of Model TS is the best among others.

In summary, the results of this experiment indicate that various urban block typologies differ quite a bit in terms of the capacities to mitigate the impact caused by ES-LEs. Among all models defined in this research, the topology of transverse slabs (TS) and longitudinal slabs (LS) might be the best choices for urban block form in waterfronts. They can highly reduce the impacts on the buildings from waves, especially for the buildings in the front. Among these two typologies, longitudinal slabs may be better than transverse slabs. Although the performances of both models in the first row are the same, more facades from the typology of transverse slabs are exposed to waves.

On the order hand, this experiment proves that the scattered layouts (Model SP, BP, STS, and SLS) may not enhance waterfronts' resiliency in comparison with compact street layouts. Relatively compact urban block form, mega blocks or courtyards (Model MP) also have a certain capacity to reduce the impact on buildings.

5.2　Results of Experiment 2

The main objective of Experiment2 is to examine the second hypothesis and reveal the optimum site coverage ratio for waterfront developments, in which simplified grid urban layouts with typical values of site coverage are designed for comparison. Six models are included in this experiment: Model A, B, C, D, E, and F; their site coverage ranges from 16% to 80%. The simulation is designed with 5000 cycles and lasts approximately 150 seconds. All facades facing the waves are registered as PartFaces to record Force-x. Because the registered PartFaces are same in terms of shape and scale, maximum Force-x is selected for comparison.

5.2.1 Model A

As shown in Figure 5.8, there are 4 buildings in Model A and its site coverage is 16%. A total of 4 facades are registered as PartFaces to record Force-x. During the simulation, 5000 values of Force-x are recorded as shown in Appendix H. The maximum and average values of Force-x on all facades are summarized in table 5.9. The maximum Force-x on every facade will be compared with other models in this experiment, while the average Force-x is for reference only.

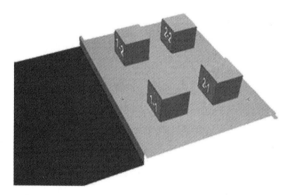

Figure 5.8   Four facades are registered as PartFaces in Model A

Maximum and average Force-x on all facades in Model A   Table 5.9

	f1-1	f1-2	f2-1	f2-2
Maximum Force-x (N)	1.33E+06	1.55E+06	6.00E+05	7.97E+05
Average Force-x (N)	4.97E+05	5.49E+05	4.26E+05	4.43E+05

5.2.2 Model B

As shown in Figure 5.9, there are 6 buildings in Model B and its site coverage is 24%. A total of 6 facades are registered as PartFaces to record Force-x. During the simulation, 5000 values of Force-x are recorded as shown in Appendix H. The maximum and average values of Force-x on all facades are summarized in table 5.10. The maximum Force-x on every facade will be compared with other models in this experiment, while the average Force-x is for reference only.

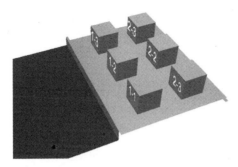

Figure 5.9　Six facades are registered as PartFaces in Model B

Maximum and average Force-x on all facades in Model B　　　　　　　　　　Table 5.10

	f1-1	f1-2	f1-3	f2-1	f2-2	f2-3
Maximum Force-x (N)	1.36E+06	1.84E+06	1.99E+06	6.40E+05	7.18E+05	7.76E+05
Average Force-x (N)	4.98E+05	5.17E+05	5.75E+05	4.35E+05	4.31E+05	4.44E+05

5.2.3　Model C

As shown in Figure 5.10, there are nine buildings in Model C and its site coverage is 36%. All nine facades facing waves are registered as PartFaces to record Force-x. The simulation results are shown in Appendix H. The maximum and average values of Force-x on all facades are summarized in Table 5.11. The maximum Force-x on every facade will be compared with other models in this experiment, while the average Force-x is for reference only.

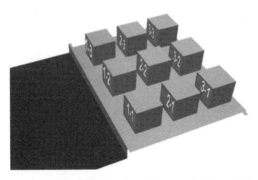

Figure 5.10　Nine facades are registered as PartFaces in Model C

Maximum and average Force-x on all facades in Model C　　　　　　　　　　Table 5.11

	f1-1	f1-2	f1-3	f2-1	f2-2	f2-3	f3-1	f3-2	f3-3
Maximum Force-x (N)	1.33E+06	1.49E+06	1.93E+06	5.52E+05	5.55E+05	6.77E+05	5.17E+05	5.13E+05	5.28E+05
Average Force-x (N)	4.95E+05	5.15E+05	6.22E+05	4.25E+05	4.35E+05	4.52E+05	4.10E+05	4.19E+05	4.28E+05

5.2.4 Model D

As shown in Figure 5.11, there are 12 buildings in Model D and its site coverage is 48%. A total of 12 facades are registered as PartFaces to record Force-x. The simulation results are shown in Appendix H. The maximum and average values of Force-x on all facades are summarized in Table 5.12. The maximum Force-x on every facade will be compared with other models in this experiment, while the average Force-x is for reference only.

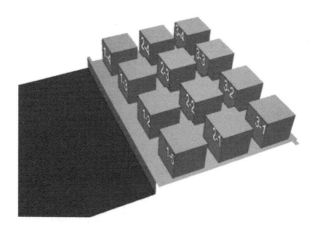

Figure 5.11 Twelve facades are registered as PartFaces in Model D

Maximum and average Force-x on all facades in Model D   Table 5.12

	f1–1	f1–2	f1–3	f1–4	f2–1	f2–2	f2–3	f2–4	f3–1	f3–2	f3–3	f3–4
Maximum Force-x (N)	1.76 E+06	1.69 E+06	1.77 E+06	2.43 E+06	7.24 E+05	7.13 E+05	8.00 E+05	9.25 E+05	5.27 E+05	5.56 E+05	6.04 E+05	6.22 E+05
Average Force-x (N)	5.77 E+05	5.47 E+05	6.03 E+05	7.20 E+05	4.50 E+05	4.53 E+05	4.70 E+05	4.86 E+05	4.13 E+05	4.16 E+05	4.27 E+05	4.35 E+05

5.2.5 Model E

As shown in Figure 5.12, there are 15 buildings in Model E and its site coverage is 60%. A total of 15 facades are registered as PartFaces. The recorded values of Force-x on PartFaces are summarized in Appendix H. In Table 5.13, the maximum and average values of Force-x on all facades are presented. The maximum Force-x on every facade will be compared with other models in this experiment, while the average Force-x is for reference only.

Figure 5.12 Fifteen facades are registered as PartFaces in Model E

Maximum and average Force-x on all facades in Model E　　　　　　　　　　　　　　Table 5.13

	f1-1	f1-2	f1-3	f1-4	f1-5	f2-1	f2-2	f2-3	f2-4	f2-5
Maximum Force-x (N)	2.12 E+06	2.13 E+06	2.57 E+06	2.84 E+06	3.35 E+06	9.53 E+05	8.64 E+05	8.80 E+05	1.01 E+06	1.11 E+06
Average Force-x (N)	6.38 E+05	6.47 E+05	6.99 E+05	7.78 E+05	8.51 E+05	4.84 E+05	4.79 E+05	4.82 E+05	4.96 E+05	5.11 E+05
	f3-1	f3-2	f3-3	f3-4	f3-5					
Maximum Force-x (N)	6.16 E+05	6.05 E+05	6.44 E+05	6.51 E+05	6.97 E+05					
Average Force-x (N)	4.22 E+05	4.21 E+05	4.22 E+05	4.30 E+05	4.39 E+05					

5.2.6 Model F

As shown in Figure 5.13, there are five long buildings in Model F and its site coverage is 80%. Only five facades are registered as PartFaces to record Force-x. The simulation results are shown in Appendix H. The maximum and average values of Force-x on all facades are summarized in Table 5.14. The maximum Force-x on every facade will be compared with other models in this experiment, while the average Force-x is for reference only.

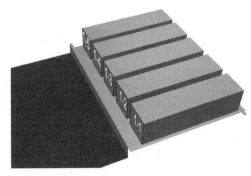

Figure 5.13 Five facades are registered as PartFaces in Model F

Maximum and average Force-x on all facades in Model F   Table 5.14

	f1-1	f1-2	f1-3	f1-4	f1-5
Maximum Force-x (N)	1.78E+06	1.82E+06	1.82E+06	2.03E+06	2.71E+06
Average Force-x (N)	5.49E+05	5.46E+05	5.73E+05	6.56E+05	7.42E+05

5.2.7 Comparison and analysis

This experiment simulates an exactly the same extreme sea level event scenario on six models (Model A, B, C, D, E, and F) respectively. With different combinations, these models are characterized as various site coverage ratios, from 16% to 80%. As shown in Table 5.15, maximum values of Force-x on all facades facing the waves in all models are summarized. Normally, with the rise of site coverage ratio, the quantity of facades on each row increases in all models except in Model F. Similar with Experiment 1, the recorded maximum Force-x reduces quite significantly from buildings in the front to the ones in the back. Therefore, the comparison will be carried out according to the relative positions of facades. In other words, the maximumvalues of Force-x in the first row, second row and third row are compared respectively. However, it is more complicated than the first experiment because the quantity of facades in the same row also varies. Taking the first row as an example, there are two facades in Model A, three facades in Model B and C, four facades in Model D, and five facades in Model E and F.

Maximum Force-x (N) of all models in experiment 2   Table 5.15

Row 1	Site cover-age	F1-1	F1-2	F1-3	F1-4	F1-5	Average of Maximum Force-x
Model A	16%	1.33E+06	1.55E+06	N/A	N/A	N/A	1.44E+06
Model B	24%	1.36E+06	1.84E+06	1.99E+06	N/A	N/A	1.73E+06
Model C	36%	1.33E+06	1.49E+06	1.93E+06	N/A	N/A	1.58E+06
Model D	48%	1.76E+06	1.69E+06	1.77E+06	2.43E+06	N/A	1.91E+06
Model E	60%	2.12E+06	2.13E+06	2.57E+06	2.84E+06	3.35E+06	2.60E+06
Model F	80%	1.78E+06	1.82E+06	1.82E+06	2.03E+06	2.71E+06	2.03E+06
Row 2	Site cover-age	f2-1	f2-2	f2-3	f2-4	f2-5	Average of Maximum Force-x

	Site cover-age	f3-1	f3-2	f3-3	f3-4	f3-5	Average of Maximum Force-x
Model A	16%	6.00E+05	7.97E+05	N/A	N/A	N/A	6.99E+05
Model B	24%	6.40E+05	7.18E+05	7.76E+05	N/A	N/A	7.11E+05
Model C	36%	5.52E+05	5.55E+05	6.77E+05	N/A	N/A	5.95E+05
Model D	48%	7.24E+05	7.13E+05	8.00E+05	9.25E+05	N/A	7.91E+05
Model E	60%	9.53E+05	8.64E+05	8.80E+05	1.01E+06	1.11E+06	9.63E+05
Model F	80%	N/A	N/A	N/A	N/A	N/A	N/A

Row 3

	Site cover-age	f3-1	f3-2	f3-3	f3-4	f3-5	Average of Maximum Force-x
Model A	16%	N/A	N/A	N/A	N/A	N/A	N/A
Model B	24%	N/A	N/A	N/A	N/A	N/A	N/A
Model C	36%	5.17E+05	5.13E+05	5.28E+05	N/A	N/A	5.19E+05
Model D	48%	5.27E+05	5.56E+05	6.04E+05	6.22E+05	N/A	5.77E+05
Model E	60%	6.16E+05	6.05E+05	6.44E+05	6.51E+05	6.97E+05	6.43E+05
Model F	80%	N/A	N/A	N/A	N/A	N/A	N/A

As shown in Chart 5.3, on most of facades in the first row, the maximum Force-x increases from Model A to B, and drops from B to C, and then increases again from C to E, and finally decreases again from E to F. A solitary exception is the change from C to D on f1-3, which is a slight decrease. The sum of all maximum Force-x divided by the total numbers of facades means the average value of maximum Force-x which could represent the overall performance of every model in the first row. As shown in the Chart 5.4, compared to model B, the maximum force on model C decreases slightly although the site coverage ratio increases. From model C to model D the values increase dramatically.

In the second row, the trend is similar with the one in first row except for Model F which is not included. Interestingly, the comparison of the maximum Force-x on f2-1 and f2-2 between Model A and C shows that although the site coverage ratio increases by 20% from A to C, the force decreases quite a lot. Since the quantity of facades varies between models, same comparison method is used again, which is the way to obtain the average value and to compare as shown in Chart 5.5. It is very clear that the overall capacity of Model C to reduce the impact is the strongest. The comparison of the third row also indicates that the impacts of ESLEs on Model C are lower than the ones on Model D and E.

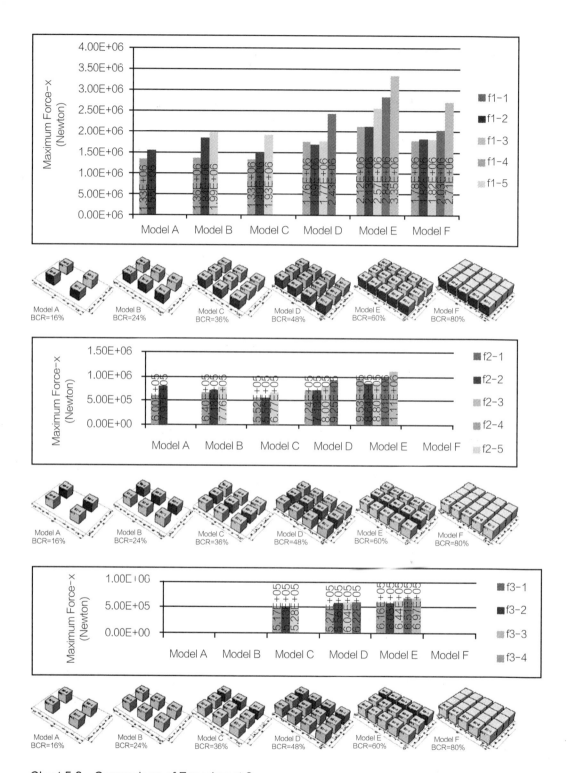

Chart 5.3　Comparison of Experiment 2

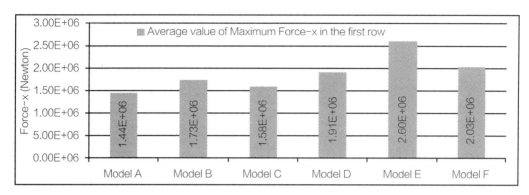

Chart 5.4 Average of maximum Force-x in the first row and first row

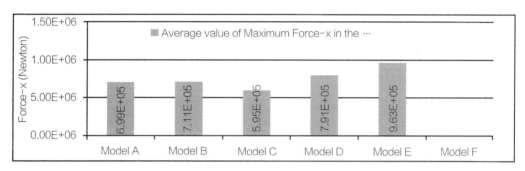

Chart 5.5 Average value of maximum Force-x in the first row and second row

More interestingly, the simulation results show that although Model F is 20% higher than Model E in terms of site coverage ratio, the impact of waves on Model F is much less than on Model E. For instance, from Model E to F, the Force-x reduces by 3.40E+05 N (16.0%) on f1-1, reduces by 3.10E+05 N (14.5%) on f1-2, reduces by 7.50E+05 N (29.1%) on f1-3, reduces by 8.10E+05 N on f1-4 (28.5%) on f1-4, reduce 6.4E+05 N (19.1%); and on average, the maximum Force-x can reduce by 21.9% (form 2.60E+06 N to 2.03E+06 N). This comparison further verifies that the block typology of longitudinal slabs is more effective than the one of small pavilions.

In summary, for this experiment, results show that Model C (site coverage ratio=36%) may be the best solution to balance coastal resiliency and land utilization ratio. Obviously, its land utilization ratio is higher thanin models A and B. Moreover, its capacity to deal with extreme sea level events is much better than thatofmodels D and E, and even slightly better than the case of model B. However, it does not indicate precisely that 36% is the only good option for all waterfront development. In addition, in this experiment, the scale of buildings is quite small and the urban fabric is quite homogeneous. Such factors might affect the results on some level. Therefore, the results of this experiment only suggest that planners can formulate the site coverage around 36% for homogeneous layouts. However, this experiment has proved the second hypothesis that there is a range of development density that can balance waterfront resiliency and land utilization efficiency for certain cases.

5.3 Results of Experiment 3

Experiment 3 was designed to compare the impact of ESLEs on models with different building depths. The third hypothesis is proposed that with other conditions being the same, increasing the depths of buildings might reduce the pressure on facades facing waves during ESLEs. Three models are designed in this experiment, namely Model G, Model H, and Model I. The model G is composed of 32 smaller buildings (width=35m, depth=25m, height=50m), while Model H is composed of 16 relatively larger buildings (width=35m, depth =60m, height=50m), and the buildings in Model I have the longest depth (width=35m, depth =135m, height=50m). The simulations are designed with 8000 cycles and last for less than 200 seconds. Only the facades with the same locations are registered as PartFaces to record Force-x. Because the area of all registered PartFaces equate with each other, maximum Force-x is selected as the variable for comparison.

5.3.1 Model G

In Model G, a total of 8 facades are registered as PartFaces to record Force-x, which are f1-1, f1-2, f1-3, f1-4, f2-1, f2-2, f2-3 and f2-4 as shown in Figure 5.14. During the simulation, 8000 values of Force-x are recoded as shown in Appendix I. The maximum and average values of Force-x on 8 PartFaces are summarized in Table 5.16. The maximum Force-x will be compared with other models in this experiment, while the average Force-x is for reference only.

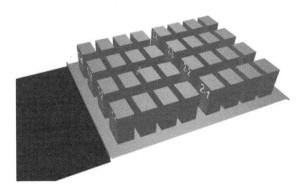

Figure 5.14 Eight registered facades in Model G

Maximum and average Force-x on registered facades in Model G Table 5.16

	f1-1	f1-2	f1-3	f1-4	f2-1	f2-2	f2-3	f2-4
Maximum Force-x (N)	4.39 E+06	5.07 E+06	6.11 E+06	5.27 E+06	1.72 E+06	1.62 E+06	1.55 E+06	1.56 E+06
Average Force-x (N)	2.20 E+06	2.23 E+06	2.17 E+06	2.21 E+06	1.49 E+06	1.48 E+06	1.47 E+06	1.47 E+06

5.3.2 Model H

In the same manner as for Model G, a total of eight facades are registered as Part-Faces to record Force-x as shown in Figure 5.15. During the simulation, 8000 values of Force-x are recoded as shown in Appendix H. The maximum and average values of Force-x on 8 registered PartFaces are summarized in Table 5.17. The maximum values Force-x will be compared with other models in this experiment, while the average Force-x is for reference only.

Figure 5.15 Eight registered PartFaces in Model H

Maximum and average Force-x on registered facades in Model H Table 5.17

	f1-1	f1-2	f1-3	f1-4	f2-1	f2-2	f2-3	f2-4
Maximum Force-x (N)	5.00 E+06	5.30 E+06	5.98 E+06	5.73 E+06	1.67 E+06	1.76 E+06	1.59 E+06	1.60 E+06
Average Force-x (N)	2.24 E+06	2.28 E+06	2.05 E+06	2.13 E+06	1.49 E+06	1.49 E+06	1.48 E+06	1.48 E+06

5.3.3 Model I

In Model I, there are also eight facades registered as PartFaces to record Force-x as shown in Figure 5.15. During the simulation, 8000 values of Force-x are recoded as shown in appendixH. The maximum and average values of Force-x on

8 registered PartFaces are summarized in Table 5.17. The maximum Force-x on these facades will be compared with other models in this experiment, while the average Force-x is for reference only.

Figure 5.16　Eight registered PartFaces in Model I

Maximum and average Force-x on registered facades in Model I　　　　Table 5.18

	f1-1	f1-2	f1-3	f1-4	f2-1	f2-2	f2-3	f2-4
Maximum Force-x (N)	4.32 E+06	5.41 E+06	5.38 E+06	5.54 E+06	1.86 E+06	1.73 E+06	1.59 E+06	1.62 E+06
Average Force-x (N)	2.26 E+06	2.18 E+06	2.02 E+06	2.06 E+06	1.49 E+06	1.49 E+06	1.48 E+06	1.47 E+06

5.3.4　Comparison and analysis

Unlike the first two experiments, the comparison of this experiment is quite straightforward since the quantity and position of registered facades are identical. Maximum values of Force-x on four facades (f1-1, f1-2, f1-3, and f1-4) in the front of the model and the ones on another four facades (f2-1, f2-2, f2-3, and f2-4) in the middle of the model will be compared respectively. Maximum values of Force-x on registered facades and average value of maximum Force-x in the front and in the middle are summarized in Table 5.19.

Maximum Force-x (N) of all models in experiment 3　　　　Table 5.19

	f1-1	f1-2	f1-3	f1-4	Average value of maximum Force-x in the front
Model G	4.39E+06	5.07E+06	6.11E+06	5.27E+06	5.21E+06
Model H	5.00E+06	5.30E+06	5.98E+06	5.73E+06	5.50E+06
Model I	4.32E+06	5.41E+06	5.38E+06	5.54E+06	5.16E+06
	f2-1	f2-2	f2-3	f2-4	Average value of maximum Force-x in the middle

Model G	1.72E+06	1.62E+06	1.55E+06	1.56E+06	1.61E+06
Model H	1.67E+06	1.76E+06	1.59E+06	1.60E+06	1.66E+06
Model I	1.86E+06	1.73E+06	1.59E+06	1.62E+06	1.70E+06

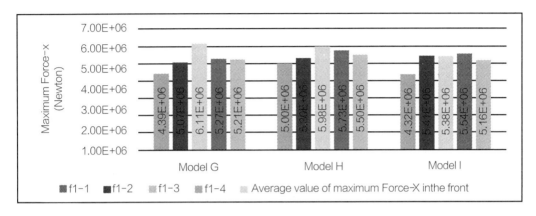

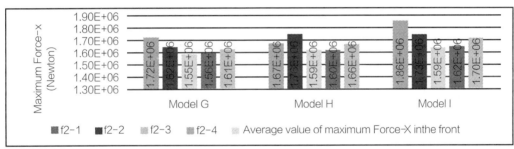

Chart 5.6   Comparison of Experiment 3

The comparison is shown in Chart 5.6. On f1-1, the highest value (5.00E+06 N) of maximum Force-x is found in Model H, while the lowest one (4.32E+06 N) is found in Model I. On f1-2, the highest one and the lowest one are respectively found in Model I (5.41E+06 N) and Model G (5.07E+06 N), but the difference (3.60E+05 N) is not as significant as the on f1-1. The most significant difference is found on f1-1 which is 7.30E+05 N between the highest value (6.11E+06 N) in Model G and the lowest value (5.38E+06 N) in Model I. On f1-4, the highest value is 5.73E+06 N in Model H and the lowest value is 5.27E+06 N in Model G. Regarding the Average value of maximum Force-x in the front, the highest value is found in Model H (5.50E+06 N) and the lowest one (5.16E+06 N) is found in Model I.

The comparison of the maximum Force-x in the middle of models shows that the facades in Model I bear more impact than other two models. However, the difference is not significant as in the front of the models. For instance, the biggest difference can be found on f2-1 between the highest value (1.86E+06 N) in Model I and the lowest value (1.67E+06 N) on Model H, which are only 1.9E+05 N. Through the comparison of average value of maximum Force-x in the middle, a clear trend can be found, which is that the impact increased slightly from Model G to Model I.

Through the comparison, a clear conclusion cannot be drawn out. Among the three models in this experiment, Model H seems to be the least effective in terms of the capacity to reduce the impact of ESLEs. The difference between the performance of Model G and Model I is not significant. Considering that there are more facades in Model G which may bear the impacts, it is believed that Model I represents the best solution among three models in this experiment. However, it is hard to say that the increased depth of buildings will definitely reduce the impact. Although this experiment cannot prove the hypothesis, it is believed that blocks' depth is an important factor that can enhance or undermine the resiliency of waterfront. Like in the second experiment, an optimum range of block's depth might be determined. In future studies more precise research should be carried out on such issue.

5.4 Results of Experiment 4

Experiment 4 intends to verify the fourth hypothesis and check the effect of changing the orientation of blocks. Four models are created as shown in the Figure 4.14: Model J, K, L, and M. Model J is modified based on Model H in Experiment 3, which has nine buildings with their long side perpendicular to the shoreline, which means $\alpha^{[1]}=0°$. In Model K, there are also nine buildings but the angle α increased to 15°, while the angle α increases even further to 30° in Model L and to 45° in Model M. In this experiment, the maximum FXPUW is considered as the variable to compare the impact due to the inequality of PartFaces being compared. The simulations are designed with 5000 to 8000 cycles and last for less than 160 seconds. Only the facades with relatively same locations are registered as PartFacesto record Force-x.

5.4.1 Model J

There are a total of are nine buildings in Model J and all buildings are identical in

---

[1] α is the angle between the long side of buildings and perpendicular of shoreline. For more information please refer to section 4.3.6, page79.

terms of shape and scale. All nine facades facing waves are registered as Part-Faces to record Force-x, which are f1-1, f1-2, f1-3, f2-1, f2-2, f2-3, f3-1, f3-2 and f3-3 as shown in Figure 5.17. During the simulation, 8000 values of Force-x are recorded as shown in Appendix J. The projected width on Y axis and the maximum values of Force-x and FXPUW on nine registered PartFaces are summarized in Table 5.20. Maximum FXPUW will be selected as a variable for comparison in this experiment.

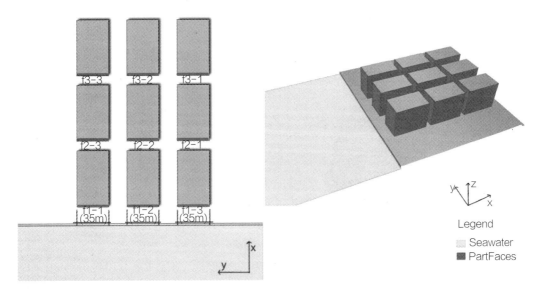

Figure 5.17 Nine registeredfacades in Model J

Maximum Force-x and FXPUW on registered facades in Model J Table 5.20

	f1-1	f1-2	f1-3	f2-1	f2-2	f2-3	f3-1	f3-2	f3-3
Projected width on Y axis (m)	35.00	35.00	35.00	35.00	35.00	35.00	35.00	35.00	35.00
Maximum Force-x (N)	3.16 E+06	3.23 E+06	3.58 E+06	1.93 E+06	1.85 E+06	1.76 E+06	1.75 E+06	1.71 E+06	1.64 E+06
Maximum FX-PUW (N/m)	9.03 E+04	9.22 E+04	1.02 E+05	5.51 E+04	5.29 E+04	5.03 E+04	5.00 E+04	4.89 E+04	4.69 E+04
Average Force-x (N)	2.22 E+06	2.19 E+06	2.13 E+06	1.61 E+06	1.61 E+06	1.55 E+06	1.54 E+06	1.53 E+06	1.51 E+06

5.4.2 Model K

In Model K, there are also nine buildings but they vary in terms of shape and scale. As shown in Figure 5.18, nine facades are registered as PartFaces to record

Force-x, including f1-1, f1-2, f1-3, f2-1, f2-2, f2-3, f3-1, f3-2 and f3-3. The projected width of registered facades on Y axis is also marked in Figure 5.18. During the simulation, 7555 values of Force-x are recoded as shown in Appendix I. The maximum values of Force-x and FXPUW on nine PartFaces are summarized in table 5.21. Maximum FXPUW will be selected as a variable for comparison in this experiment.

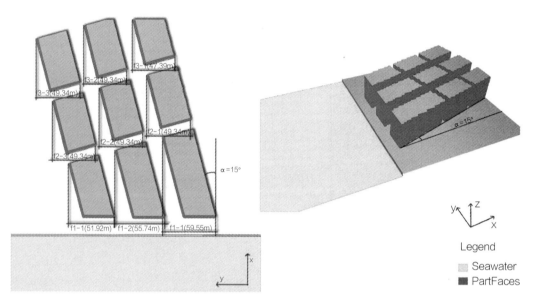

Figure 5.18   Nine registeredfacades in Model K

Maximum Force-x and FXPUW on registered facades in Model K   Table 5.21

	f1-1	f1-2	f1-3	f2-1	f2-2	f2-3	f3-1	f3-2	f3-3
Projected width on Y axis (m)	59.55	55.74	51.92	49.34	49.34	49.34	47.39	49.34	49.34
Maximum Force-x (N)	3.63 E+06	3.29 E+06	3.61 E+06	1.92 E+06	1.96 E+06	1.98 E+06	1.74 E+06	1.78 E+06	2.04 E+06
MFXPW (N/m)	6.09 E+04	5.90 E+04	6.95 E+04	3.89 E+04	3.97 E+04	4.01 E+04	36.7 E+04	3.61 E+04	4.13 E+04
AverageForce-x (N)	2.49 E+06	2.24 E+06	2.05 E+06	1.57 E+06	1.58 E+06	1.58 E+06	1.58 E+06	1.54 E+06	1.60 E+06

5.4.3   Model L

As shown in Figure 5.19, there are 10 buildings in Model L and they vary in terms of shape and scale. A total of nine facades are registered as PartFaces to record Force-x, which are f1-1, f1-2, f1-3, f2-1, f2-2, f2-3, f3-1, f3-2, and f3-3. The projected width of registered facades on Y axis is also marked in Figure 5.19. During the simulation, 6241 values of Force-x are recoded as shown in Appendix J. The maximum values of Force-x and FXPUW on nine PartFaces are summarized in Table 5.22. Maximum FXPUW will be selected as a variable for comparison in this experiment.

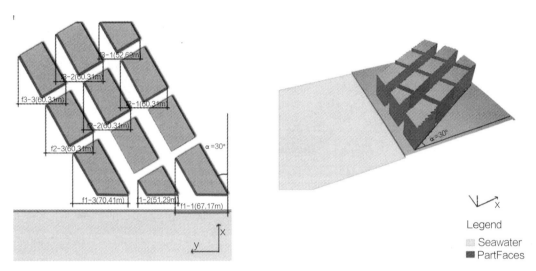

Figure 5.19　Nine registeredfacades in Model L

Maximum Force-x and FXPUW on registered facades in Model L　　　　　Table 5.22

	f1-1	f1-2	f1-3	f2-1	f2-2	f2-3	f3-1	f3-2	f3-3
Projected width on Y axis (m)	67.17	51.29	70.41	60.31	60.31	60.31	52.69	60.31	60.31
Maximum Force-x (N)	3.63E+06	3.29E+06	3.61E+06	1.92E+06	1.96E+06	1.98E+06	1.75E+06	1.78E+06	2.05E+06
MFXPW (N/m)	5.40E+04	6.41E+04	5.13E+04	3.18E+04	3.25E+04	3.28E+04	2.95E+04	2.95E+04	3.40E+04
Average Force-x (N)	2.44E+06	2.20E+06	2.00E+06	1.55E+06	1.56E+06	1.55E+06	1.57E+06	1.52E+06	1.57E+06

5.4.4　Model M

As shown in Figure 5.20, there are nine buildings in Model M and they vary in terms of shape and scale. A total of nine facades are registered as PartFacesto record Force-x, which are f1-1, f1-2, f1-3, f2-1, f2-2, f3-1, f3-2 and f3-3. The projected width of registered facades on Y axis is also marked in Figure 5-20. During the simulation, 5530 values of Force-x are recoded as shown in Appendix J.

The maximum values of Force-x and FXPUW on nine PartFaces are summarized in Table 5.23. Maximum FXPUW will be selected as a variable for comparison in this experiment.

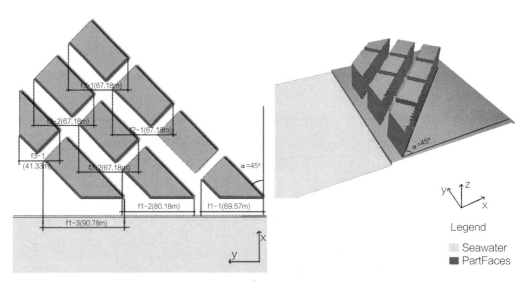

Figure 5.20　Eight registeredfacades in Model M

Maximum Force-x and FXPUW on registered facades in Model M　　　　　　　Table 5.23

	f1-1	f1-2	f1-3	f2-1	f2-2	f3-1	f3-2	f3-3
Projected width on Y axis (m)	69.57	80.18	90.78	67.18	67.18	67.18	67.18	41.33
Maximum Force-x (N)	5.73 E+06	3.64 E+06	5.26 E+06	2.19 E+06	2.68 E+06	1.52 E+06	2.35 E+06	1.60 E+06
MFXPW (N/m)	8.23 E+04	4.54 E+04	5.79 E+04	3.26 E+04	3.99 E+04	2.26 E+04	3.50 E+04	3.87 E+04
Average Force-x (N)	2.69 E+06	2.06 E+06	2.08 E+06	1.24 E+06	1.41 E+06	1.10 E+06	1.25 E+06	3.78 E+05

5.4.5　Comparison and analysis

The fourth experiment simulates an exactly same extreme sea level events scenario on four models (Model J, K, L, and M) that are characterized by the angel α. With the increase of the angel α, the layouts of models are changed as shown in Figure 4.14. Except for Model M which has eight registered facades, others have nine PartFaces to record the Force-x during the simulation. Due to the dissimilarity of buildings' quantity and position, the comparison has to be carried out based on relative positions of registered facades, which means that the impacts in the first row, the second row and the third row will be compared respectively. Meanwhile, the area registered facades vary widely, which means it is impossible to compare the Force-x directly. It is because that the impact caused by ESLEs may rise with the increase of facades' area. Therefore, as discussed in the previous chapter, maximum FXPUW is defined as the variable of this experiment. As shown in Table 5.24, the maximum values of FXPUW on all registered facades and the average of maximum FXPUW in each row are summarized.

Maximum FXPUW (N/m) on all PartFaces and the average for each row   Table 5.24

	α	Maximum FXPUW			Average of maximum FXPUW in the first row
		f1-1	f1-2	f1-3	
Model J	0°	9.03E+04	9.22E+04	1.02E+05	9.48E+04
Model K	15°	6.09E+04	5.90E+04	6.95E+04	6.31E+04
Model L	30°	5.40E+04	6.41E+04	5.13E+04	5.65E+04
Model M	45°	8.23E+04	4.54E+04	5.79E+04	6.19E+04
	α	Maximum FXPUW			Average of maximum FXPUW in the second row
		f2-1	f2-2	f2-3	
Model J	0°	5.51E+04	5.29E+04	5.03E+04	5.28E+04
Model K	15°	3.89E+04	3.97E+04	4.01E+04	3.96E+04
Model L	30°	3.18E+04	3.25E+04	3.28E+04	3.24E+04
Model M	45°	3.26E+04	3.99E+04	N/A	3.63E+04
	α	Maximum FXPUW			Average of maximum FXPUW in the third row
		f3-1	f3-2	f3-3	
Model J	0°	5.00E+04	4.89E+04	4.69E+04	4.86E+04
Model K	15°	3.67E+04	3.61E+04	4.13E+04	3.87E+04
Model L	30°	2.95E+04	2.95E+04	3.40E+04	3.10E+04
Model M	45°	2.26E+04	3.50E+04	3.87E+04	3.21E+04

Chart 5.6 shows the comparison between three models. In the first row, a same trend can be seen on f1-1, f1-3 and the average. It shows that from Model J to K (α increase from 0° to 15°), maximum FXPUW decreases dramatically. For instance, on f1-1, it drops by 32.6% from Model J (9.03E+04 N/m) to K (6.09E+04 N/m). From K to L (α increases from 15° to 30°), the impact also decreases slightly (11.3% on f1-1, 26.2% on f1-3 and 10.5% on average). When the angle α increases from 30° to 45° (from Model L to M), maximum FXPUW ascends. However, on f1-2, the trend is more complicated. The highest value (9.22E+04 N/m) is also found in Model J and it also decreases significantly (36.0%) from Model J to K. Unlike on other facades, the second highest value (6.41E+04 N/m) is found

in Model L. Based on the comparison, it is quite clear that Model J is the least effective design in terms of the capacity to adapt to ESLEs. Taking the average value of FXPUW in the first row as the main consideration, it is believed that Model L has the strongest capacity to respond to ESLEs.

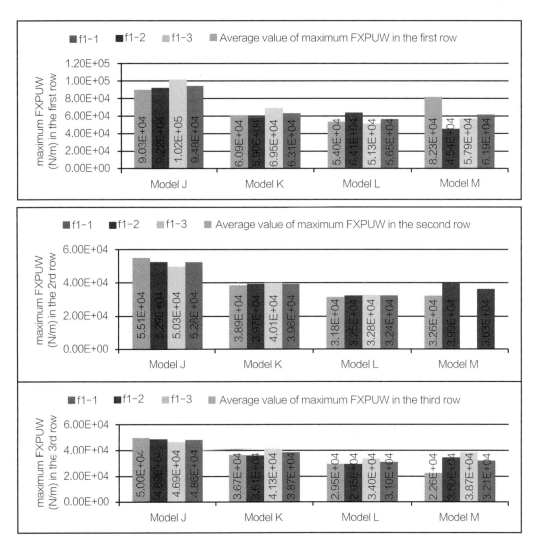

Chart 5.7 Comparison for Experiment 4

The comparison in the second row is clearer than the one in the first row. On all facades and the average value, it shows a relatively significant decrease from Model J to K, another decrease from Model to L, and a slight increase from Model L to M except f2-3 in Model M which is not included. In the third row, except for the fact that maximum FXPUW decreases from Model L to M on f3-3, a similar trend could also be noticed.

Based on the comparison above, it is quite clear that Model L with the angle $\alpha=30°$ has the most effective layout to deal with ESLEs. It also indicates that changing the orientation of blocks or buildings can highly reduce the impact from ESLEs. Taking the comparison of the first row as an example, maximum FXPUW could be reduced at most 49.7% (form 1.02E+05 N/m in Model J to 5.13E+04 N/m Model L on f1-3). However, when the angle α increases from 30° to 45°, the impacts on facades show a rising trend, which might be because more area will be exposed to waves when α is greater or equal to 45°.

In summary, this experiment has verified the fourth hypothesis proposed in this thesis, which is that changing block orientations can help to mitigate the damage caused by ESLEs. However, it is necessary to keep α less than 45° when this adaptation is implemented.

5.5 Results of Experiment 5

Experiment 5 is designed to verify the fifth hypothesis in this thesis, which is to check the performance of different types of U-shaped blocks and to identify the one with the best capacity to respond to ESLEs. In Figure 4.15, four models representing typical U-shaped blocks are created, namely Model N, O, P, and Q. Among them, Model N is a complete U-shaped block; Model O is a block composed by three individual buildings; Model P is composed by two perfectly symmetric buildings; and Model Q is composed by an L-shaped building and a rectangle building.

The simulations are designed with 6000 cycles and last for about 260 seconds. All facades facing the waves are registered as PartFaces to record Force-x. The registered facades on both sides in all models are identical, but the ones in the middle of the blocks are different in shape and scale. The simulation results of this experiment are summarized in appendix K. Therefore, similar with Experiment 4, the maximum FXPUW is selected as the variable to compare the impact of ESLEs on the facades.

5.5.1 Model N

There is only one building in Model N and a total of three facades registered as PartFacesto record Force-x, including f1, f2 and f3 as shown in Figure 5.21. The width of registered facades facing waves is also marked in Figure 5.21 and summarized in Table 5.25. The maximum values of Force-x and FXPUW on three registered PartFaces are also summarized in Table 5.25. The maximum FXPUW is the main variable to assess the impact on facades.

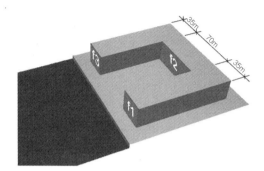

Figure 5.21   Three registered facades in Model N

Maximum Force-x and FXPUW on registered facades in Model N   Table 5.25

	f1	f2	f3
Projected width of facade on Y axis (m)	35	70	35
Maximum Force-x (N)	4.91E+06	9.14E+06	4.73E+06
Maximum FXPUW (N/m)	1.40E+05	1.31E+05	1.35E+05
Average Force-x (N)	2.00E+06	3.77E+06	1.88E+06

5.5.2   Model O

There are three individual buildings in Model O and a total of three facades registered as PartFacesto record Force-x, including f1, f2 and f3 as shown in Figure 5.23. The width of registered facades facing waves is also marked in Figure 5.23 and summarized in Table 5.26. The maximum values of Force-x and FXPUW on three registered PartFaces are also summarized in Table 5.26. The maximum FXPUW is the main variable to assess the impact on facades.

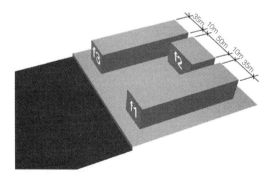

Figure 5.22   Three registered facades in Model O

Maximum Force-x and FXPUW on registered facades in Model O   Table 5.26

	f1	f2	f3
Projected width on Y axis (m)	35	50	35
Maximum Force-x (N)	4.26E+06	5.60E+06	4.20E+06
Maximum FXPUW (N/m)	1.22E+05	1.12E+05	1.20E+05
Average Force-x (N)	1.86E+06	2.37E+06	1.78E+06

5.5.3 Model P

There are two buildings in Model P and a total of three facades are registered as PartFaces to record Force-x (f1, f2 and f3) as shown in Figure 5.23. In this model, two unconnected facades are registered as one PartFace, namely f2. The width of registered facades facing waves is also marked in Figure 5.23 and summarized in Table 5.27. The maximum values of Force-x and FXPUW on three registered PartFaces are also summarized in Table 5.27. The maximum FXPUW is the main variable used to assess the impact on facades for this experiment.

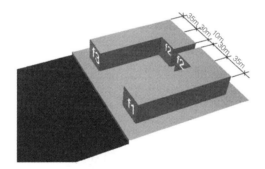

Figure 5.23   Three registered facades in Model P

Maximum Force-x and FXPUW on registered facades in Model P   Table 5.27

	f1	f2	f3
Projected width on Y axis (m)	35	60	35
Maximum Force-x(N)	4.31E+06	7.21E+06	4.46E+06
Maximum FXPUW (N/m)	1.23E+05	1.20E+05	1.27E+05
Average Force-x(N)	1.78E+06	2.90E+06	1.78E+06

5.5.4 Model Q

There are two buildings in Model Q and a total of three facades are registered as PartFaces to record Force-x, including f1, f2 and f3 as shown in Figure 5.24. The width of registered facades facing waves is also marked in Figure 5.24. and summarized in Table 5.28. The maximum values of Force-x and FXPUW on three registered PartFaces are also summarized in Table 5.28. The maximum FXPUW is the main variable to assess the impact on facades for this experiment.

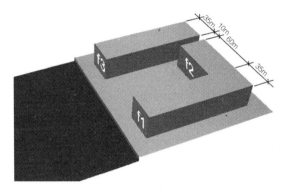

Figure 5.24   Three registered facades in Model Q

Maximum Force-x and FXPUW on registered facades in Model Q   Table 5.28

	f1	f2	f3
Projected width on Y axis (m)	35	60	35
Maximum Force-x (N)	4.38E+06	7.84E+06	4.12E+06
Maximum FXPUW (N/m)	1.25E+05	1.30E+05	1.17E+05
Average Force-x (N)	1.84E+06	2.98E+06	1.81E+06

5.5.5 Comparison and analysis

In this experiment, four types of U-shaped blocks are tested under a same scenario of extreme sea level events, which include a regular U-shaped block and three modified U-shaped block tested under a same extreme sea level event simulation. In every model, three facades are defined as PartFaces to record Force-x. The analysis is based on the comparison of maximum FXPUW on PartFaces that are summarized in Table 5.29.

As shown in Chart 5.8, on the one hand, the highest values on the three facades (f1, f2, and f3) are all found in model N, which suggests that regular U-shaped block (Model N) is not a good urban form for waterfront development. On the other hand, almost all the lowest values can be found in Model O which represents the U-shaped block are separated into three parts as shown in Figure 5.22. It suggests that by changing the block form from Model N to O the impacts on facades can be highly reduced. For instance, maximum FXPUW can be respectively reduced by 12.9% on f1 and 11.1% on f3; and for the facade in the middle, f2, it can be reduced by 14.5%.

Maximum FXPUW (N/m) on all registered facades Table 5.29

	f1	f2	f3
Model N	1.40E+05	1.31E+05	1.35E+05
Model O	1.22E+05	1.12E+05	1.20E+05
Model P	1.23E+05	1.20E+05	1.27E+05
Model Q	1.25E+05	1.30E+05	1.17E+05

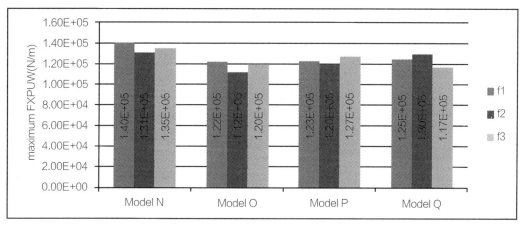

Chart 5.8 Comparison for experiment 5

In addition, the comparison also shows that Model P and Model Q can reduce the impacts. For example, comparing to Model N with Model P, the impact can be slightly reduced by 12.0% on f1, 8.4% on f2 and 5.9% on f3. The comparison between Model N and Q shows that almost no difference can be found on f2. Nonetheless, Model Q can considerably reduce the impacts on f1 (10.7%) and f3 (13.3%).

The results verify the last hypothesis of this research and indicate that in waterfront developments, it is better to avoid choosing U-shaped building form. If a U-shaped building is required, necessary modifications are required to mitigate the impact of ESLEs. Three modifications provided in this research are able to reduce the impact from ESLEs and separating U-shaped buildings into three parts (Model O) is the most effective way to enhance waterfronts' resiliency.

5.6 Chapter summary

In this chapter, the results of five experiments have been presented. Based on the results, the comparisons and analyses have been conducted in order to examine the hypotheses proposed in previous chapters. Among five hypotheses, four of them are proved to be true and only one may need more in-depth studies. According to the analyses, four methods have been proved to be effective in response to ESLEs, including (1) transforming urban blocks' typologies, (2) adjusting the site coverage ratio of waterfront developments, (3) changing the blocks' orientations, and (4) modifying U-shaped blocks. However, regarding the third hypothesis, it is difficult to prove that the increased depth of buildings will definitely reduce the impact, but it is still believed that blocks' depth is an important factor that can enhance or undermine the resiliency of waterfronts. In future studies, more precise experiments should be designed in order to reveal the specific details.

Based on the above presented experiments' results, the research question can be answered by stating that urban form based adaptations are effective to reduce the impact of ESLEs on waterfront buildings. First, the impact of ESLEs on the buildings closet to seawater can be reduced by more than 10% if the urban block typology is transformed from scattered ones (such as the typologies represented by Model SP and BP) to the compact ones (such as mega blocks, courtyards and street slabs). Secondly, adjusting the site coverage ratio into an optimum range for waterfronts also helps to enhance their resiliency without compromising coastal land utilization efficiency. This research discovers that for a waterfront composed by small scaled buildings, the optimum site coverage ratio is 36%. Thirdly, changing the buildings' orientation is also an effective method to enhance waterfronts' capacity to respond to ESLEs. For instance, if the angle between the shoreline and facades is increased from 0° to 30°, the impact of ESLEs can be reduced by 49.7% at the most in some cases. Last but not least, this research also proves that separating U-shaped buildings can effectively reduce the impact of ESLEs. For example, as showed in experiment 5, if a U-shaped building is divided into three individual parts, the impact of ESLEs can decrease by more than 10%.

In summary, the experiments in this research have theoretically proved four hypotheses. The application of these meso-scale urban form-based adaptations into real waterfront developments is still an important issue since the ultimate research purpose is to upgrade urban design prentices for waterfront development. Therefore, in the next chapter, an application case study will be addressed with the intention of integrating these adaptations within a waterfront development proposal and discovering their capacities to reduce the impact of ESLEs in real cases.

Chapter 6 Case study and of Singapore

In the last chapter, five hypotheses have been examined. The results of experiments indicate that four hypotheses are verified while the third hypothesis cannot be proved since it needs more accurate parametric studies in order to reveal the change pattern of impact on facades when the blocks' depth is increased. In this chapter, a case study is carried out based on a waterfront development proposal in Singapore in order to further prove hypotheses' applicability. The proposal for the waterfront development will be modified and integrated in meso-scale urban form-based adaptations. Using the same methodology, an extreme sea level event is will be simulated on both the original and modified proposals. During the simulation, impacts on selected facades will be recorded and compared with the intention to identify the design to have a better capacity to reduce the damage caused by ES-LEs.

In the beginning of this chapter, the impacts of sea level rise and relative extreme events on Singapore are introduced, which is followed by the discussion of Singapore's response strategies. Subsequently, the background information of the chosen case is presented. Thereafter, the case study will be illustrated through Figures, tables and charts. By the end of this chapter, a discussion will be elaborated based on the results of the case study application.

6.1 Challenges for waterfront development in Singapore

As discussed in the second chapter, the sea level rise and climate change are deeply threatening waterfront developments in coastal cities. In terms of sea level rise and extreme sea level events, Singapore's circumstance is better than other coastal citiesfrom the Asian Pacific region due to its location that is protected by periphery islands. For instance, after assessing a 72-year tidal record of Hong Kong and factors such as estuarine backwater effects and long-term geological subsidence, the results suggest that a 30 cm rise in relative sea level at the mouth of the estuary is possible by 2030(Z Huang et al., 2004). The increase of sea level around Pearl River Delta would destroy most of the reclaimed land of Hong Kong, Macau and Shenzhen(Z Huang et al., 2004). However, current research has shown that Singapore is also quite vulnerable and exposed to climate change and sea level rise. The government has stated in a national report, National Climate Change Strategy, that coastal land loss and increasing flood risk are two major impacts that should be addressed in the near future (NCCS, 2008). For waterfront developments in Singapore, coastal land loss and frequent extreme sea level events are two important issues which require integrated response strategies. In this section, the two major threats will be addressed respectively.

6.1.1 Coastal land loss

With no doubt, to many coastal areas, the biggest threat caused by the rising sea is coastal land loss and most island states are suffering from it. Singapore is one of the densest cities in the world. Currently the population in Singapore is more than 5 million, and the target population of Singapore Concept Plan 2011 is expected to reach 6.0 million(URA, 2010). To accommodate a population like this within a small island is an enormous challenge, and accordingly many new settlements on the reclaimed land are required. Nevertheless, without proper protections, these coastal settlements will be endangered greatly if the sea level will continue to rise in this century.

In a national report named Singapore National Climate Change Strategy(NCCS, 2008) it is very clearly stated that sea level rise could lead to coastal erosion and land loss, especially as Singapore has a relatively flat coastline. It also declares that East Coast Park, SungeiBuloh, PasirRis Park, West Coast Park, and Sentosa are the most vulnerable locations (ibid).

Wong(1992) estimated the possible impacts of one-meter sea level rise based on observations of highest tide event on 9th February 1974. This tide level is 3.9m above the Chart Datum, which is highest tide in record. It is 1.1m higher than the current Mean High Water Spring [1](MHWS, 2.8m) and its effect could be assumed as similar as one meter sea level rise. According to Wong's research, one-meter sea level rise could cause evident erosion on the beaches, especially on the reclaimed land on the Southeast, the West and North coast (Wong, 1992).

Chan(1999) adopted the Bruun Rule as the fundamental theory to study the in-

---

[1] The Mean High Water Spring (MHWS) has an average value for highest levels the spring tides reach over a period of time. For more information, please refer to http://en.wikipedia.org/wiki/Mean_high_water_spring.

undation of coast in Singapore. Ten beaches in Singapore and periphery islands were chosen, including both natural beaches and artificial beaches. The future sea level rise projection released by IPCC in 1995 was adopted. According to this projection, by 2100 the sea level will rise by 15cm to 95cm based on different scenarios. The shoreline retreats by 2100 have been calculated as shown in table 6.1. According to this research, Changi and Sentosa rank as the most vulnerable beaches in Singapore. The shorelines retreat of these two locations will be respectively 28.5m and 38m (Chan, 1999). In conclusion, it is predicted that the shoreline retreats of manmade beaches is generally worse than for the natural beaches (ibid). Unfortunately, most beaches in Singapore are artificial ones.

Similarly, Ng and Mendelsohn (2005) chose ten coastal sites in Singapore to estimate the land loss under the circumstances of different sea level change scenarios. The selected ten sites represent the major developed land uses areas in Singapore. Parts of the results are show in the Table 6.2. Given an assumption of 0.86m sea level rise by 2100, the land loss over time is simulated. According to this study, the CBD area is the most vulnerable location along the Singapore's coastal line; and there will be 2.19 square kilometers land loss due to the erosion (W. S. Ng & Mendelsohn, 2005)

Projection of shoreline retreats by 2100 in Singapore
Source: Chan, Y. M. (1999), edited by author. Table 6.1

Beaches in Singapore	Description	Projected sea level rise (the most optimistic estimate and the worst scenario) by 2100 (m)	Shoreline retreats (m) by 2100
Pasir Ris 1	Manmade beach between breakwaters, facing north	0.95	10.184
		0.15	1.608
Pasir Ris 2		0.95	14.706
		0.15	2.322
Pasir Ris 3		0.95	9.31
		0.15	1.47
Pasir Ris 4		0.95	7.125
		0.15	1.125
Changi	Natural beach facing north	0.95	28.5
		0.15	4.5
East Coast	Manmade beach facing south	0.95	95
		0.15	15

Beaches in Singapore	Description	Projected sea level rise (the most optimistic estimate and the worst scenario) by 2100 (m)	Shoreline retreats (m) by 2100
NoordinBeache (PulauUbin)	Natural beach, North island, facing north	0.95	57
		0.15	9.7
Kusu	Natural beach, Southern island, north and south facing beaches	0.95	9.5
		0.15	1.5
Siloso Beach (Sentosa)	Manmade beach, Southern island, facing south	0.95	38
		0.15	6
Lazarus Island Beach	Natural beach, Southern island, north and south facing beaches	0.95	9.5
		0.15	1.5

Time series of inundated land areas (sq km) for each coastal site under the 0.86m sea level rise scenario. Source: Ng, W. S. and R. Mendelsohn (2005) p. 209.　　　Table 6.2

Year	Marine				Pulau Keppel					
	Loyang	Parade	CBD	Sentosa	Bukom	Harbor	Jurong	Tuas	Kranji	Woodlands
2000	0.05	0.09	0.20	0.01	0.20	0.02	0.11	0.10	0.07	0.05
2010	0.10	0.18	0.40	0.02	0.39	0.05	0.22	0.20	0.14	0.11
2020	0.14	0.27	0.60	0.03	0.59	0.07	0.33	0.30	0.21	0.16
2030	0.19	0.36	0.80	0.04	0.78	0.09	0.44	0.40	0.28	0.21
2040	0.24	0.45	1.00	0.05	0.98	0.12	0.55	0.50	0.35	0.26
2050	0.29	0.54	1.20	0.06	1.17	0.14	0.66	0.60	0.42	0.32
2060	0.34	0.63	1.40	0.07	1.37	0.16	0.77	0.70	0.49	0.37
2070	0.39	0.72	1.60	0.08	1.56	0.18	0.88	0.80	0.56	0.42
2080	0.43	0.81	1.80	0.09	1.76	0.21	0.99	0.89	0.63	0.47
2090	0.48	0.90	2.00	0.10	1.95	0.23	1.10	0.99	0.70	0.53
2100	0.53	0.99	2.19	0.11	2.15	0.25	1.24	1.09	0.78	0.58

6.1.2 Increasing risk of flooding and storm

It has been discussed in the second chapter that together with global warming, sea level rise may cause more extreme sea level events in terms of both frequency and intensity. As an island state in Southeast Asia, Singapore has to be aware of the increasing risks of coastal disasters. Economic Cooperation and Development (OECD) had published a study demonstrating that port cities will endure the intense effects of climate change, including sea-level rise, violent storms, flooding and land subsidence, if the prediction of global warming is correct (Zhuang,

2010). In this comparative research, Singapore was given a relatively low risk ranking, which is No. 79 among the world's 136 major port cities. Nevertheless, as a port city whose economy largely depends on maritime activities and coastal area densely populated, sea level rise and extreme sea level events are attracting growing concerns.

In addition, Wong (1992) also mentions that an increase in sea level could undermine the drainage system and cause coastal flooding as the channels cannot flow out to the sea. Likewise, Singapore's National Climate Change Strategy has addressed that "a higher sea level rise makes it more difficult for rainwater drain into the sea" and "can aggravate inland flooding during storm surges and rainstorms" (NCCS, 2008, p. 8).

Protected by periphery islands, Singapore is thought to be safe regarding storm surges and tsunamis. However, recent research shows that Singapore is not totally immune from these extreme sea level events. It is found that it takes about 12 hours for the tsunami waves generated at Manila Trench to arrive at Singapore coastal waters; the time interval between two peaks is about five hours; the maximum water level rise in Singapore water is about 0.8 m; and the maximum velocity associated with the tsunami waves is about 0.5 m/s (Z. Huang et al., 2009).Tkalich et al (2009)also conclude that the primary causative factor of sea level anomalies in Singapore Strait is the wind over the South China Sea; and the anomalies could be classified as storm surges with 0.8 meters tides. In addition, Kolomiets et al (2010) discovered that for extreme cases, the maximum sea surface may reach a height up to 1 meter in the Singapore Strait.

Based on the current situation, these events could not cause catastrophic damage to Singapore's waterfront development; and some of them could not even be noticed. However, considering 1 to 2 meters sea level rises, the threats could not be underestimated in the future. It is addressed that Singapore is possible to be hit by tsunamis, and Changi Airport and the manmade island of Jurong, which houses a sprawling petrochemical complex, are very vulnerable and even low waves can cause damages to them (Ho, 2008).

6.2 Response of Singapore

Based on the discussion above, the potential impacts are inevitable if Singapore government takes no action to respond. In another word, Singapore still stands chances to have adaptive options implemented to enhance waterfront resilience. According to the Third Assessment Report of IPCC, the most serious considerations for small island states is whether they will have an adequate potential to adapt to sea level rise within their own country boundaries(J. J. McCarthy, 2001). Singapore is a wealthy and well-organized country, which provides the possibility for financial and technical support.

In fact, Singapore government takes climate change very seriously. In 2010, The National Climate Change Secretariat (NCCS) was set up as a dedicated agency under the Prime Minister's Office, whose responsibility is to manage Singapore's domestic and international affairs on climate change issues. NCCS has promised to develop a national climate change strategy and to ensure that Singapore is prepared for future impacts of climate changes. The resilience workgroup in NCCS, which is headed by the Ministry of the Environment and Water Resources and the Ministry of the National Development, is responsible for Singapore's vulnerability to the impacts of climate change and aims to develop appropriate adaptation plans. Moreover, other governmental agencies such as National Environment Agency(NEA), Urban Redevelopment Authority (URA), Building & Construction Authority (BCA),and Public Utilities Board (PUB),are also devoted to responding to climate changes. Meanwhile, research institutes are involved into the study of climate change in order to provide solutions, such as Institute of Catastrophe Risk Management (ICRM) at Nanyang Technological University (NTU), the Risk Management Institute (RMI), and Tropical Marine Science Institute (TMSI) in National University of Singapore (NUS).

Nevertheless, most of policies carried out by Singapore government are focusing on mitigations and reducing greenhouse gas emission, such as the energy-effective program. As Wong (1992) addresses it, although Singapore is aware of the global climate change , it has not announced any research or strategy specifically focused on sea level rise. Only a few details of current strategies to deal with sea level rise and related coastal disasters can be extracted from existing policies.

According to Strategies for Adaptation to Sea Level Rise (Dronkers et al., 1990), there are four main approaches to deal with the sea level rise, which are coastal risk management, retreat, protection and accommodation (adaptation). Based on these four categories, current policies will be examined and the capacity of Singapore to cope with sea level rise and extreme events could be assessed.

6.2.1　Coastal risk management

In 2007, the National Environment Agency commissioned the Vulnerability Study (V Study), which is the first research project to assess Singapore's vulnerability to climate change. With the objective of projecting the effect and impact of global climate change, the V Study was undertaken by TMSI in NUS. Using the methodology developed by IPCC, it projects that "the average daily temperature in Singapore may increase between 2.7℃ to 4.2℃ by 2100, and that the mean sea level around Singapore may rise by between 24cm to 65cm by 2100" [1].

Another assessment called Risk Map Study was commissioned by the Building & Construction Authority. With the intent to better identify the specific coastal areas at risk of inundation and the potential damage associated with it, the entire Singapore coastline will be covered and several site topographical surveys could be employed; the results are expected to help the development of coastal protection strategies against inundation caused by sea level changes(NCCS, 2012).

6.2.2　Retreat

For a small island country with a dense population, retreat from coastal areas is not a practical option. On the contrary, in order to accommodate the growing population, Singapore has been continuously putting efforts on coastal developments and land reclamation. Up to now, a percentage of 10% to 20% of the original land was reclaimed from the sea around Singapore (Qing, 2005). Although they are facing the rising sea and changing climate, coastal areas are in fact the focus of development and attract quite a lot of attention and investments due to advantaged landscape resource.Currently, Singapore had launched several schemes for seafront.For instance, under the Singapore Concept Plan 2011, it is feasible to consolidate a comprehensiveport at Tuas, which could free up a current major port (TanjongPagar), and provide land for new development. Singapore's Economic Strategic Committee (EMC) had released their report in 2010 and appealed to redevelop Tanjong Pagar area as the extension of CBD (ESC, 2010). Obviously, it can change the image of the city's waterfront and attract more investments. However, in terms of future sea level rise and extreme events, this area could be incredibly vulnerable as it was built up as a port several decades ago when sea

---

[1]　Please refer to http://www.nea.gov.sg/cms/ar2011/safeguard-5.html

level rise was not a main concern.

6.2.3 Protection

As discussed in the third chapter, protection strategies involve defensive measures and other activities to protect areas against inundation, tidal flooding, effects of waves on infrastructure, shore erosion, salinity intrusion, and the loss of natural resources (Dronkers et al., 1990). In Singapore, several government agencies, including the Housing & Development Board (HDB), Building and Construction Authority (BCA), and JTC Corporation have been involved in coastal protection and engineering work. For example, a Coastal and Project Management Department has been set up in BCA to focus on coastal protection and adaptation issues(NCCS, 2012). Nowadays, except of few natural beaches, hard walls or stone embankments could be found in 70%-80% of Singapore's coastal areas, which could help to protect against coastal erosion.

Meanwhile, the soft coastal protection practice also grows with the governmental support, such as the projects of beach nourishment and mangrove plantation. And according to National Climate Change Strategy 2012, local institutions will be embarked on to advance these areas and new technology to make Singapore's coasts more resilient(NCCS, 2012).

Sea level rise and global warming are without doubts destroying the coastal ecosystem as it is documented in the National Climate Change Strategy 2012 that: "mangroves along the north-eastern coast of the north-eastern coast of the island of PulauTekong were being scoured as a result of coastline erosion corals, which require sunlight, may not be able to grow upwards quickly enough to keep pace with rising sea levels. In addition, a 1° C to 2° C rise in sea water temperature will lead to coral bleaching, which occurred in 1998 and 2006, andkilled at least 16% of coral reefs worldwide." (NCCS, 2012, p. 85). Led by National Parks Board (NParks) a multidisciplinary team is set up to achieve the restoration of mangrove habitats, and the environmentally sustainable soft engineering solutions through combination of rocks,mangrove plants and biodegradable rings (ibid).

6.2.4 Accommodation

Regarding accommodation, most of the Singapore government's efforts are concentrated in promoting drainage systems. To increase Singapore's flood resilience, the Ministry of the Environment and Water Resources (MEWR) has reviewed the design of drainage system and flood protection measures. Meanwhile, PUB, Singapore's national water agency,is also involved in the improvement of flood management and the forecasting system through modeling simulation, widening catchment and expanding the drainage systems. In order to cope with the rising sea level and with intensifying rainfall, 20 of the drainage upgrading projectsare expected to be accomplished within five years (ibid). Meanwhile, concerning sea level rise, since 1991, all new land reclamation projects must be built 1.25m above the highest recorded tide level. And in 2011, in order to adapt long-term sea level rise, BCA required land level of new projects must be raised by another additional 1m (ibid).

6.2.5 Section summary

In Singapore, basically the government takes climate change and sea level rise quite seriously. Related governmental agencies and research institutions have been involved in creating response strategies, and have achieved satisfying results, such as the vulnerability research,improvement of drainage system and enhancement of protections. Nevertheless, currently neither researchers, nor the policy makers have paid enough attention to the increasing risk of extreme sea level events, such as storm surge floods or tsunamis. Considering the advantages, such as geographical position and relatively shallow waters of Straits of Malacca, the risk that Singapore is affected by a fierce storm surge or tsunami is relatively low. Yet,if the future rise of sea level increases and the global climate further deteriorates, the probability of such events might increase. Therefore, to totally ignore the risk of further sea level rise will lead to serious consequences.

Currently, Singapore's urban planning system has not formulated a series of strategies to respond to sea level rise and ESLEs for waterfront developments. In 2011, the senior Executive Engineer of Building and Construction Authority, Mr.Ho Chai Teck was interviewed by author. He admitted that urban planning authorities of Singapore were aware of the possible impacts of climate change and sea level rise on waterfront developments but they did not intend to adapt the design methods and regulations to such challenges in the discipline of physical planning except raising coastal land elevation. He also addressed that authorities encourage scholars to develop adaptations for waterfront developments in Singapore.

In summary, current Singapore's response strategies include finishing the risk assessment and management, enhancing protections, upgrading infrastructures and raising coastal land revelation. Overall, it is believed the next step of Singapore's response strategies is to upgrade the urban planning and design system and reveal which kind of urban form is more resilient in waterfronts in the near future. Therefore, this thesis takes one waterfront development in Singapore as case study in order to verify the feasibility of the meso-scale urban form based on adaptations that were examined in the last chapter. The case is chosen from the proposal of Marina South, and its details are introduced in next section.

6.3 Waterfront development in Marina South

As shown in Figure 6.1, Marina South is located in the Southern waterfront of Singapore and within Marina Bay, which is being master planned by Urban Redevelopment Authority as the next stage of urban development. As shown in Figure 6.2, URA has set aside 60ha of land between Garden by the Bay and the Straits of Singapore, where will "raise some 11,000 homes, with a mix of commercial, hotel and community facilities". However, as a piece of land reclaimed from sea, the potential threats from the rising sea level are not highlighted by URA.

Marina Bay is the center of Singapore and transforms this city into a vibrant, global city. Several major landmarks have already been earmarked in this district, including the Marina Bay Sands Integrated Resort, and the 100ha Gardens by the Bay. Marina South is going to be the extension of Marina Bay, which is just minutes away from the city center. In the Concept Plan 2001, URA's long term plan for Singapore's development over the next 40 to 50 years, has created more city living options for Singaporeans (URA, 2001). Nowadays, city living with splendid views of waterfront is the lifestyle which Singaporeans would like to choose. Marina South will provide attractive city living among lush greenery (Garden by the Bay) and beautiful waterscape, with panoramic views across city's signature skyline (URA, 2007).

In order to attractive this area and generate innovative concepts for next-generation living environment, the URA and Singapore Institute of Architects (SIA) are jointly organizing a Design Ideas Competition for the Marina South Residential District. A total of 30 entries were submitted, and the concepts of green spaces, highrise development, social space, etc. were emphasized by competitors. However, neither the organizers, nor the designers have highlighted the potential threat from the rising sea and extremes events.

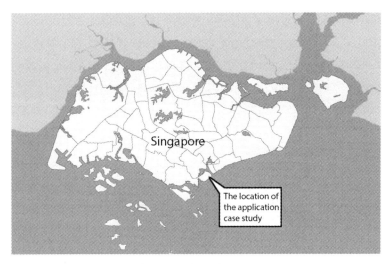

Figure 6.1   Location of case study (Marina South)

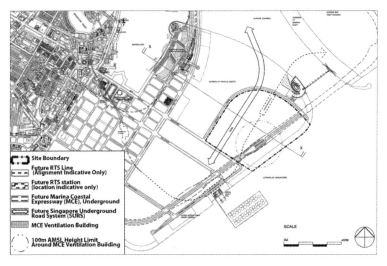

Figure 6.2   The planned area for Marina South

Currently, the development of Marina South is in the concept planning stage. As shown in Figure 6.3, a draft design has been sketched by URA in order to assess its feasibility and environmental performance, including transportation, ventilation, energy demand, and so on. Unfortunately, its capacity to respond to ESLEs is not included in the assessment. It can be seen from Figure 6.3, in this proposal, that a large section of this district will face the open sea directly. Without adaptations, serious damage might be inevitable. Under this circumstance, this research chooses the central part of Marina South as the study area to carry out an application test. The selected area is shown in Figure 6.4 and 6.5.

In this test, the original proposal (Design 1) will be modified according to the results of last chapter and a new proposal (Design 2) integrated with meso-scale urban form-based adaptations will be sketched. An extreme sea level event will be simulated on both designs. Using the same methodology, maximum FXPUW on the selected facades will be compared so that the capacity of adaptations can be examined.

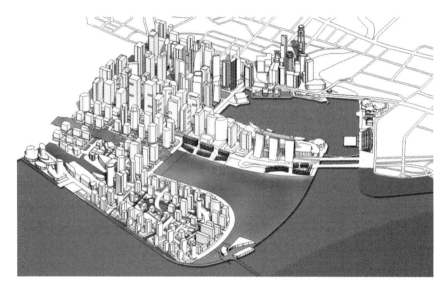

Figure 6.3   Draft design for Marina South

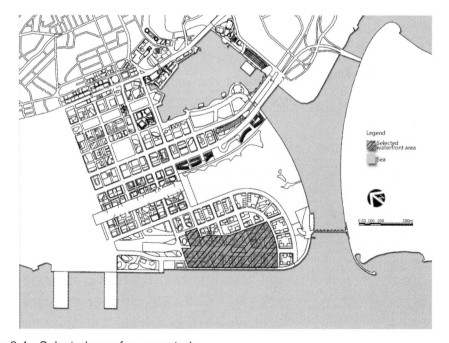

Figure 6.4   Selected area for case study

6.4 Application case study

6.4.1 Analysis and modification

Figure 6.5 shows the original design of selected area, which is named as Design 1. The selected area is about 2.93E+06 m² and its site coverage ratio is 32.05%. In order to facilitate the description and comparison, the design is divided into 12 blocks as shown in Figure 6.5. Among them, Block 1, 2, 3, 4, 5, 6 and 7 are in the front of select area, while Block 8, 9, 10, and 11 are in the back. Block 12 is in the center, which is the ventilation building for underground Marian Coastal Expressway. Design 1 is analyzed in four aspectsof physical forms, including site coverage ratio, block typologies, block orientations, and different types of U-shaped blocks, which are four verified hypotheses. Accordingly, the modification is provided based on the analyses.

1) Site coverage

It is mentioned, that the site coverage ratio is 32.1%. As proved in Chapter 5, the optimum value for site coverage ratio is around 36%. This hypothesis is proved through theoretical models, in which building are homogeneous distributed. Therefore, the applicable condition for this hypothesis is that buildings must be homogeneously arranged. Fortunately, the original design roughly satisfies such condition as shown in Figure 6.5. Therefore, regarding this aspect, the suggestion for modification is to slightly increase the site coverage ratio. Design 2 is the modified proposal as shown in Figure 6.7. After modification, the site coverage ratio increases to 33.3%.

2) Block topology

In Design 1, some of the urban blocks could be considered as semi-open courtyards, such as Block 1, 2, 3, 4, 5, 6, and 9. According to the classification of urban block typologiesin this research, this kind of typology is a combination of mega pavilions/courtyards (MP), transverse slabs (TS), and short transvers slabs (SLS). As proved in chapter 5, both MP and TS are good block forms in terms of the capacity to respond to ELSEs. However, the best typology has been proved to be the typology of longitudinal slabs (Model LS). Therefore, it is suggested that some blocks could be transformed into LS. To be specific, as shown in Figure 6.7, two buildings in the east wing of Block 1 are connected together in order to transform the SLS into LS. Meanwhile, the front part of Block 2 is separated into 3 individual

Figure 6.5 Plan of Design 1

buildings in order to turn it into a typology of LS. A similar method is applied on block 9 as shown in Figure 6.6. In addition, the central part of Block 4 is removed and two buildings in the east wing are connected so that it becomes LS block.

Block 7, 8, and 11 are more difficult to be identified in terms of block typology. For instance, on one hand, Block 7 can be identified as big pavilion (BP) according to the classification in Experiment 1 if it is considered as one block; however, on the other hand, if it is considered as four individual blocks, each building in Block 7 can be identified as a mega block (MP). As discussed in Chapter 5, the capacity of MP is better than BP. It is assumed that the impact of ESLEs might be reduced if all buildings in these blocks could be connected together and form a whole. Nonetheless, it is to be noted that such modification may not possible because it may seriously affect these buildings' indoor environments, spatial organization and open spaces. Meanwhile, such modification may also break some regulations about fire safety and minimum distance between buildings. Therefore, no changes will be made on Block 7, 8 and 11 in terms of urban block typologies.

Block 10 is a typical longitudinal slab, which is proved to be the best typology; therefore, there will be no modification on it either. Block 12 is the ventilation building for underground expressway, whose block typology can be identified as the one in between MP and TS. It is unnecessary and impossible to modify the typology due to its special function.

3) Block orientation

This thesis has proved that changing the blocks' orientation so that they are not directly perpendicular to the shoreline can also enhance the waterfronts' resiliency. Therefore, to rotate the blocks within the range of possibilities is also an effective method to modify the Design 1. As shown in Figure 6.5, in the original design, quite a lot of buildings and facades in some blocks have been skewed at a certain angle ($\alpha=21°$) with the shoreline, including Block 6, 7, 8, 9, 10, 11, and 12. Although this outcome is probably not driven by the concern of ESLEs, it is believed that such layout will help to enhance the buildings' capability to respond to extreme events. Therefore, as shown in Figure 6.7, with the intention to reduce the impact, other blocks, such as Block 3, 4, 5 and 6, are also skewed 21° northwards during the modification. Meanwhile, slight changes are made on Block 8, 10 and 11 in order to make more front facades not to face the wave directly. Besides, the west facade of Block 12 is also skewed at the same angle as shown in Figure 6.7.

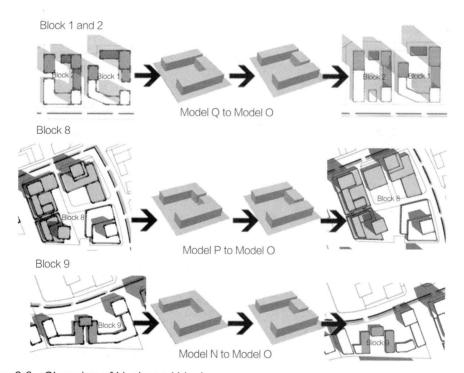

Figure 6.6 Changing of U-shaped blocks

4) U-shaped block

As discussed in the previous chapter, several U-shaped blocks (Model N, P and

Q in Experiment 5) are not a good block forms to deal with ESLEs and transforming them into a U-shape block that is composed by three individual parts (Model O in Experiment 5) can highly reduce the impact from ESLEs. In Design 1, several blocks can be distinguished due to their spatial characteristic of U-shape, including Block 1, 2, 8 and 9. But these U-shaped blocks identified in Design 1 may not be the best typologies in terms of response to ESLEs, and require changes. For instance, the back parts of Block 1 and 2 can be obviously identified as U-shaped block represented by Model Q. Meanwhile, the entire Block 8 can be considered as U-shaped form like Model P. Besides, the middle part of Block 9 is a typical U-shaped form represented by Model N. As shown in Figure 6.6, all U-shaped blocks found in Design 1 are transformed in the process of modification.

Figure 6.7 Plan of Design 2

After modification, a new waterfront design (Design 2) with integrated effective adaptations is sketched as shown in Figure 6.7. Overall, a total of four meso-scale urban form-based adaptations that have been successfully applied in this case, which include (1) adjusting the site coverage ratio of waterfront developments, (2) transforming urban blocks' typologies, (3) changing the blocks' orientations, and (4) modifying U-shaped blocks. Comparing Design 1 and Design 2, it can be seen that the overall layouts do not vary greatly. After the application of adaptations, new designed blocks can harmoniously coexist with other parts. More importantly, although quite a lot of changes have been made, it is believed that original functions can basically fit into the new design. In the next section, the same extreme sea level event scenario will be simulated on both designs and the impacts on selected facades will be compared.

6.4.2 Simulation design

In order to compare the performances of the two designs, an extreme sea level event is simulated on both of them. The method to simulate the extreme sea level events is the same with other experiments in this research. The wave height is set as 12 meters and water period is 20 seconds. The specific setting for extreme sea level events can be found in appendix K. Actually, as previously stated, Singapore may not be affected by an extreme event like this under the current circumstances. However, the extreme sea level event simulated in this case study is assumed to be happening in the future with the assumption that sea level will continue to rise and climate will continue to change.

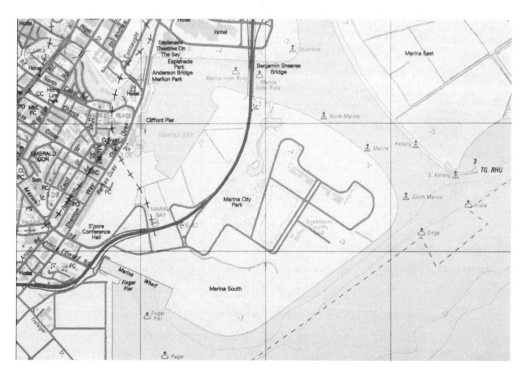

Figure 6.8 Current land elevations in Marina South. Source: Singapore's map published by Mapping Unit, Ministry of Defense, Singapore in 2005, edited by author

Due to the fact that the land elevation of Marina South is currently classified information this study calculates it based on the Singapore's map published by Mapping Unit, Ministry of Defense, Singapore in 2005. As shown in Figure 6.9, the land elevation of study area is 3 meters above sea level in 2005. As mentioned in section 6.1.4, the land elevations of new developments on reclaimed land are required to be elevated up to 2 meters. Assuming that sea level will rise 1 meter by the end of this century, the final land elevation for this case is calculated as 4 meters (3m+2m−1m) above sea level. Acceding to the original design, the distance between buildings and shoreline is at most 32 meters and at least 25 meters. The beach is designed at an incline with the slope of 2/5 as shown in Figure 6.9. The sea level is designed as being 4 meters below the land elevation of study area.

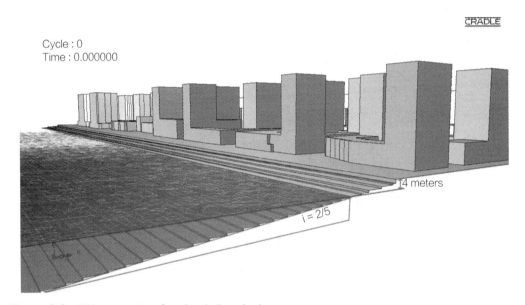

Figure 6.9   Urban context for simulation design

As shown in Figure 6.9 and 6.10, in both Design 1 and Design 2, almost all facades face waves are registered as PartFaces to record the impact of the extreme sea level event on them. The detailed information about all PartFaces is shown in Figure 6.12-6.19, including the number of the facades and their projected width on Y-axis. There are a total of 45 facades registered in 12 blocks. In these Figures, all PartFaces are highlighted by a red line. With the illustration provided by the Figures, the difference between Design 1 and Design 2 could be clearly presented.

The simulation lasts for about 160 seconds and more than 8000 values of Forex are recorded on the selected facades. The maximum FXPUW represents the impact on facades which is chosen for comparison since the area of facades may change after modifications.

Figure 6.10　All registered PartFaces to record Force-X in Design 1

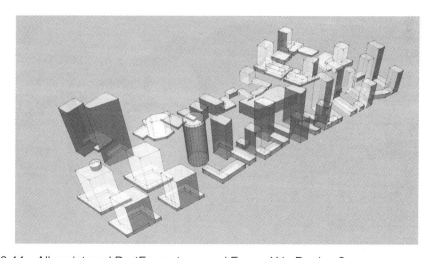

Figure 6.11　All registered PartFaces to record Force-X in Design 2

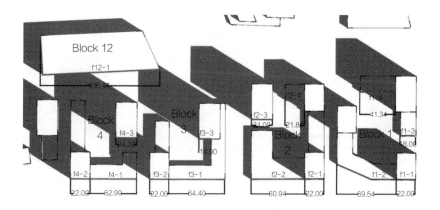

Figure 6.12　Registered PartFaces of Block 1, 2, 3, 4, and 12 in Design1

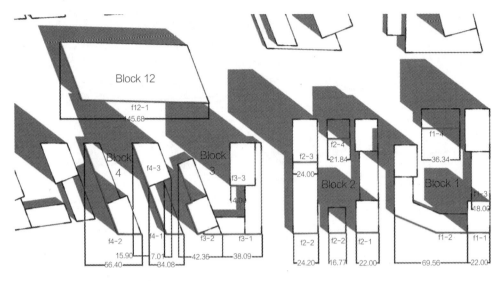

Figure 6.13 Registered PartFaces of Block 1, 2, 3, 4, and 12 in Design2

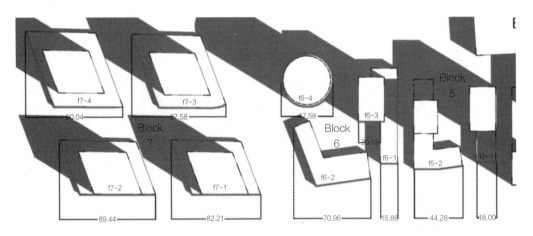

Figure 6.14 Registered PartFaces of Block 5, 6, and 7 in Design1

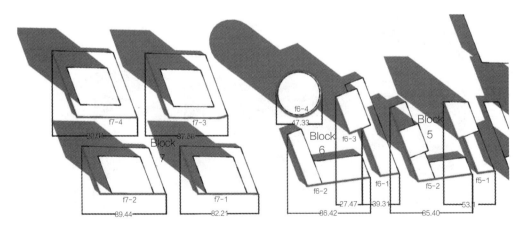

Figure 6.15 Registered PartFaces of Block 5, 6, and 7 in Design2

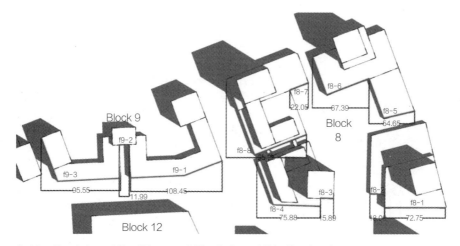

Figure 6.16   Registered PartFaces of Block 8 and 9 in Design1

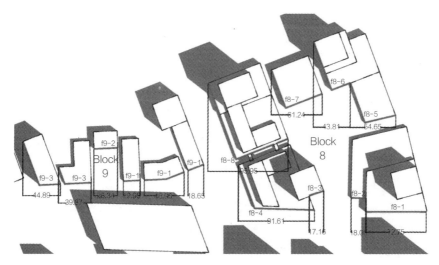

Figure 6.17   Registered PartFaces of Block 8 and 9 in Design2

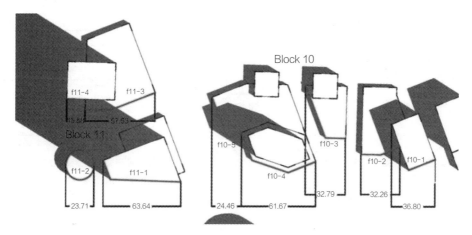

Figure 6.18   Registered Partfaces of Block 10 and 11 in Design1

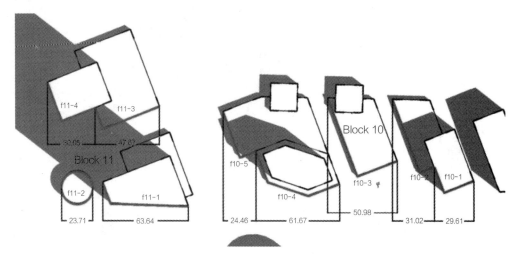

Figure 6.19　Registered PartFaces of Block 10 and 11 in Design2

6.4.3　Simulation and comparison

The simulation results of Design 1 and Design 2 are respectively shown in Table 6.4 and Table 6.5, in which the values of maximum FXPUW all PartFaces are calculated for comparison.

Chart 6.1 shows the comparison of Block 1 and 2 between Design 1 and Design 2. On f1-1 and f1-2, only a slight decrease can be found after modification. However, comparing with Design 1, obvious declines can be noticed on f1-3 and 1-4 in Design 2. There is no impact on f1-3, which could be explained by the fact that the facade is covered by the building in front of it. The decrease found on f1-4 indicates that changing U-shaped buildings from the type of Model Q to Model O is effective to reduce the impact of ESLEs.

Simulation results of Design 1　　　　　　　　　　　　　　　　　　　　　　　Table 6.3

	Block 1				Block 2			
	f1-1	f1-2	f1-3	f1-4	f2-1	f2-2	f2-3	f2-4
Width on Y-axis (m)	22.00	69.54	18.00	41.34	22.00	60.94	24.00	21.84
Maxi. Force-x (N)	3.00E+07	7.90E+07	3.34E+06	8.84E+06	3.14E+07	9.98E+07	2.75E+06	6.51E+06
Max. FXPUW (N/m)	1.36E+06	1.14E+06	1.86E+05	2.14E+05	1.43E+06	1.64E+06	1.15E+05	2.98E+05

	Block 3			Block 4			Block 5	
	f3-1	f3-2	f3-3	f4-1	f4-2	f4-3	f5-1	f5-2
Width on Y-axis (m)	64.40	22.00	14.00	52.99	22.00	24.00	18.00	44.28
Max. Force-x (N)	6.01E+07	1.70E+07	9.19E+05	3.98E+07	1.63E+07	2.19E+06	9.41E+06	2.21E+07
Max. FXPUW (N/m)	9.33E+05	7.73E+05	6.56E+04	7.51E+05	7.41E+05	9.13E+04	5.23E+05	4.99E+05
	Block 6				Block 7			
	f6-1	f6-2	f6-3	f6-4	f7-1	f7-2	f7-3	f7-4
Width on Y-axis (m)	15.88	70.96	20.12	47.58	82.21	89.44	87.58	90.04
Max. Force-x (N)	8.15E+06	3.04E+07	4.51E+06	2.42E+06	3.81E+07	3.98E+07	5.08E+06	7.44E+06
Max. FXPUW (N/m)	5.13E+05	4.28E+05	2.24E+05	5.09E+04	4.63E+05	4.45E+05	5.80E+04	8.26E+04
	Block 8							
	f8-1	f8-2	f8-3	f8-4	f8-5	f8-6	f8-7	f8-8
Width on Y-axis (m)	72.75	18.00	15.89	75.88	54.65	67.39	22.05	95.95
Max. Force-x (N)	1.12E+07	4.97E+06	2.46E+06	1.05E+07	1.97E+06	5.81E+06	1.56E+06	3.73E+06
Max. FXPUW (N/m)	1.54E+05	2.76E+05	1.55E+05	1.38E+05	3.60E+04	8.62E+04	7.07E+04	3.89E+04
	Block 9			Block 10				
	f9-1	f9-2	f9-3	f10-1	f10-2	f10-3	f10-4	f10-5
Width on Y-axis (m)	108.45	11.99	95.55	29.60	32.26	32.79	61.67	24.46
Max. Force-x (N)	6.60E+06	9.23E+05	4.22E+06	1.58E+06	1.13E+06	8.98E+05	1.57E+06	8.00E+05
Max. FXPUW (N/m)	6.09E+04	7.70E+04	4.42E+04	5.34E+04	3.50E+04	2.74E+04	2.55E+04	3.27E+04

	Block 11				Block 12				
	f11-1	f11-2	f11-3	f11-4	f12-1				
Width on Y-axis (m)	63.64	23.71	57.63	15.58	136.46				
Max. Force-x (N)	2.47E+06	5.97E+05	7.30E+05	9.08E+05	1.12E+07				
Max. FXPUW (N/m)	3.88E+04	2.52E+04	1.27E+04	5.83E+04	8.21E+04				

Simulation results of Design 2   Table 6.4

	Block 1				Block 2			
	f1-1	f1-2	f1-3	f1-4	f2-1	f2-2	f2-3	f2-4
Width on Y-axis (m)	22.00	69.56	18.00	36.34	22.00	40.97	24.00	21.84
Max. Force-x (N)	2.91E+07	7.82E+07	0	2.14E+06	2.90E+07	6.12E+07	4.11E+05	1.53E+07
Max. FXPUW (N/m)	1.32E+06	1.12E+06	0	5.89E+04	1.32E+06	1.49E+06	1.71E+04	7.00E+05

	Block 3			Block 4			Block 5	
	f3-1	f3-2	f3-3	f4-1	f4-2	f4-3	f5-1	f5-2
Width on Y-axis (m)	38.09	42.36	14.00	34.08	56.40	22.91	53.11	85.40
Max. Force-x (N)	3.45E+07	3.17E+07	9.53E+05	1.45E+07	2.02E+07	4.20E+06	1.48E+07	3.55E+07
Max. FXPUW (N/m)	9.05E+05	7.48E+05	6.81E+04	4.25E+05	3.58E+05	1.83E+05	2.79E+05	4.16E+05

	Block 6				Block 7			
	f6-1	f6-2	f6-3	f6-4	f7-1	f7-2	f7-3	f7-4
Width on Y-axis (m)	61.4	86.42	3.62	47.33	82.21	89.44	87.58	90.04
Max. Force-x (N)	1.23E+07	4.26E+07	1.29E+06	1.78E+06	3.61E+07	3.93E+07	4.24E+06	7.34E+06

Max. FXPUW (N/m)	3.13E+05	4.92E+05	4.69E+04	3.76E+04	4.39E+05	3.99E+05	4.84E+04	8.12E+04
	Block 8							
	f8-1	f8-2	f8-3	f8-4	f8-5	f8-6	f8-7	f8-8
Width on Y-axis (m)	72.75	18.00	17.16	91.61	54.65	43.81	61.24	95.95
Max. Force-x (N)	1.34E+07	5.90E+06	2.74E+06	1.87E+07	2.38E+06	3.70E+06	4.17E+06	2.97E+06
Max. FXPUW (N/m)	1.84E+05	3.28E+05	1.60E+05	2.04E+05	4.35E+04	8.45E+04	6.80E+04	3.11E+04
	Block 9			Block 10				
	f9-1	f9-2	f9-3	f10-1	f10-2	f10-3	f10-4	f10-5
Width on Y-axis (m)	88.77	29.29	82.46	29.61	31.02	50.98	61.67	24.46
Max. Force-x (N)	5.25E+06	1.22E+06	3.48E+06	1.80E+06	7.47E+05	1.33E+06	1.31E+06	6.60E+05
Max. FXPUW (N/m)	5.91E+04	4.17E+04	4.22E+04	6.08E+04	2.41E+04	2.61E+04	2.12E+04	2.70E+04
	Block 11				Block 12			
	f11-1	f11-2	f11-3	f11-4	f12-1			
Width on Y-axis (m)	63.64	23.71	47.67	36.05	145.60			
Max. Force-x (N)	2.48E+06	4.82E+05	6.34E+05	1.17E+06	1.25E+07			
Max. FXPUW (N/m)	3.91E+04	2.03E+04	1.33E+04	3.24E+04	8.55E+04			

In Block 2, after integration with adaptations, the impacts on f2-1 and f2-2 respectively reduced by 7.7% and 9.1% and a significant decrease (85.2%) can found on f2-3, which indicates the modification is successful. However, the maximum FXPUW significantly increases on f2-4, from 2.98E+05 N/m to 7.00E+05 N/m. It is believed that the protection for the buildings at the back of the blocks might lose when the buildings in the front are separated and let waves go through. It indicates that such modification method might increase the impact on the facades at the back. Therefore, this method should be carried out carefully only in the case where the buildings in the back can be compromised.

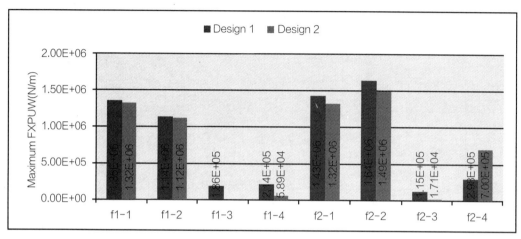

Chart 6.1　Comparison for Block 1 and 2

Chart 6.2 shows the comparison on Block 3, 4 and 5 between Design 1 and Design 2. Because no major changes have been made on Block 3, only slight decreases can be found on f3-1 and f3-2 as well as a slight increase on f1-3.

However, on Block 4, major changes can be noticed. For instance, maximum FXPUW reduces from 7.51E+05 N/m to 4.25E+05 N/m on f4-1 and from 7.41E+05 N/m to 3.58E+05 N/m on f4-2. This result indicates that the adaptations of changing blocks' orientation and transforming block typology from TS to LS are quite effective. Similar with f2-4, a significant increase also can be found on f4-3 which is the facade in the back of the block. In this case, the increase on f4-3 (9.17E+04 N/m) is so slight that it might be negligible compared with the decrease on f4-1(3.26E+05 N/m) and f4-2 (3.83E+05 N/m).

The impact on Block 5 is also highly reduced after the adaptation of changing ordinations is applied on it. As shown in chart 6.2, maximum FXPUW reduces by 46.7% (from 5.23E+05 N/m to 2.79E+05 N/m) on f5-1 and by 19.8% (from 4.9916E+05 N/m to 4.16E+05 N/m). The significant decreases of maximum FXPUW on Block 4 and 5 suggest that changing block's orientation is an effective adaptation to respond to ESLEs.

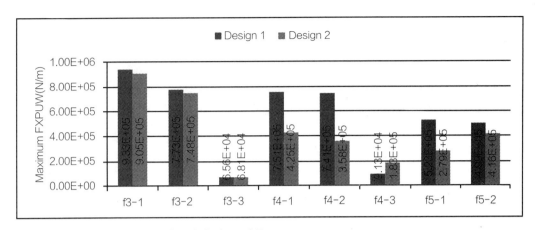

Chart 6.2 Comparison for Block 3, 4, and 5

In Block 6, maximum FXPUW reduces significantly on f6-1 and f6-3 which might be because of the adaptation of changing blocks' orientation. A slight increase can be found on f6-2 which is barely modified. On f6-4, only a slight decrease can be noticed. Block 7 is the only block without any modification. Accordingly, only a small amount of decline of maximum FXPUW can be found on four registered PartFaces.

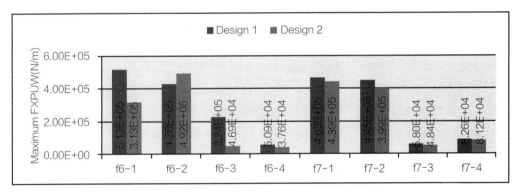

Chart 6.3 Comparison for Block 6 and 7

Chart 6.4 shows the comparison for Block 8, which is the biggest block. A total of eight facades are registered as PartFaces. The comparison between Design 1 and Design 2 shows that on one hand the impact on facades in the front is magnified, and on the other hand the impact on the facades in the back is reduced slightly after modification. One possible explanation for such phenomenon is that waves' energy will act on the buildings at the back of study area if it does not act on the buildings at the front, which are integrated with effective adaptations. In other words, when the waves travel through Block 1 and 2, some impact reduced by adaptations might act on Block 8. It is noticed that a similar phenomenon could be found in the Block 12. As it is discussed, the adaptations' ultimate target is to reduce the impact on the buildings located at the interface of land and water, which might bear the most significant impact caused by ESLEs. Therefore, it is still meaningful to put adaptations in use although it might increase the impact on the buildings at the back of waterfronts.

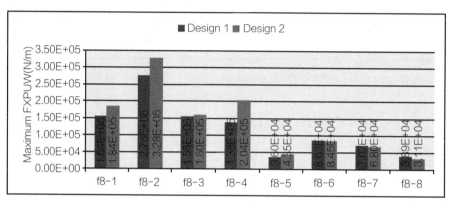

Chart 6.4　Comparison for Block 8

The comparison for Block 9 and 10 (Chart 6.4) shows that most of the impacts decrease slightly after modification except f9-2, f10-1 and f10-2. The impact on f10-1 increases by 12.2% (7.40E+04 N/m) while the impacts on f9-2 and f10-2 decrease quite significantly. For instance, the maximum FXPUW decreases from 7.70E+04 N/m in to 4.17E+04 N/m, which is a decline of 45.8%. As shown in Figure 6.6, the modification for f9-2 is the modifying of U-shaped building from Model N to Model O, which means it is a feasible and effective adaptation. Meanwhile, the maximum FXPUW on f10-2 is also reasonably reduced by 31.1%, which indicates that applying the adaptation of changing orientations is effective on f10-2.

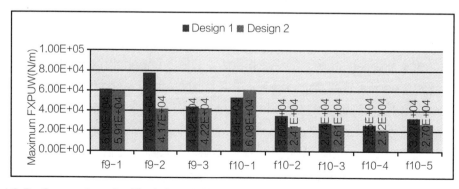

Chart 6.5　Comparison for Block 9 and 10

The comparison for Block 11 and 12 is shown in chart 6.6. In Block 11 and 12, most of the values of maximum FXPUW on facades change slightly after modification. The most significant alteration can be found on f11-4, which is a decline of 2.59E+04 N/m (44.4%) and indicates again that the adaptation of changing orientation can reduce the impact of ESLEs. A slight ascent can be found on f12-1, which is possibly because more impact might act on f12-1 when the impact is highly reduced on the blocks (Block 3 and 4) in front of Block 12. Meanwhile, the impact might also be reduced somehow by the adaptation applied on Block 12 so that only a slight increase (4.1%) can be noticed.

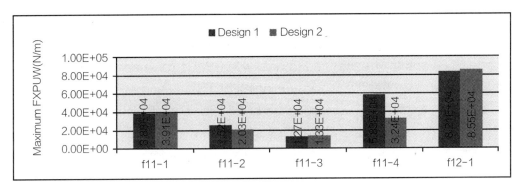

Chart 6.6   Comparison for Block 11 and 12

6.5   Discussion

In the previous chapters, it has been proved that four meso-scale urban form-based adaptations in response to ESLEs are theoretically effective, including (1) adjusting the site coverage ratio of waterfront developments, (2) transforming urban blocks' typologies, (3) changing the blocks' orientations, and (4) modifying U-shaped blocks. In this chapter, after addressing the waterfront development in Singapore and its challenges from extreme sea level events, an application case study is discussed. In this case study, four adaptations have been applied through simulation into a design proposal for the central area of Marina South waterfront development.

A conclusion can be drawn that these meso-scale urban form based adaptations can be successfully applied in waterfront developments. As shown in Figure 6.5 and 6.7, although the original urban fabric is partly changed, it is still possible to fit the designed functions in the modified proposal. Most importantly, the comparison of the simulations' results shows that the impact can be reduced by the application of adaptations.

The first adaptation, adjusting the site coverage ratio, cannot be directly proved to be effective by this case study. Actually, the original design is quite close to be optimum value that is proved to be 36% by Experiment 2. Although the site coverage ratio has been slightly increased from 32.1% to 33.3% after the modification, it is hard to judge that the decline of impact is due to the increase of site coverage ratio. In the future, more case studies should be designed to check the effect of increasing site coverage ratios on waterfront development proposals.

The second adaptation applied in this case study is transforming urban blocks' typologies. In chapter 4, seven typologies have been defined (SP, BP, MP, TS, STS, LS, and SLS) and their capacities to deal with ESLEs have been compared. The comparison shows that the block typologies of TS, LS and MP are much better than others in terms of impact reduction of ESLEs. Therefore, in this case study, the typologies of Block 1, 2 and 4 are transformed, which help to reduce the impact of ESLEs, especially on Block 4. Nevertheless, when the impact on the buildings in the front is reduced by this adaptation, more impact than expected may act on the buildings in the back. Therefore, this adaptation should be carefully applied when the buildings in the back could be compromised to a certain extent.

The third adaptation is changing the orientation of blocks or buildings. It has been proved by the case study that the impact of ESLEs can be considerably reduced if the orientation of buildings or urban blocks is not perpendicular to the shoreline. In this case study, this adaptation is the most widely used, which is applied on Block 3, 4, 5 and 6 and some buildings or facades in Block 8, 10, and 11. The results have shown that this adaptation is quite effective on most of the blocks. Nonetheless, there is still a concern about this adaptation since the changing of orientation might affect the indoor and outdoor space of buildings. Accordingly, the rotation of orientation must be controlled within the permitted range.

The fourth adaptation is to modify U-shaped blocks. In Chapter 5, it has been concluded that different U-shaped blocks vary widely in terms of their response to ESLEs. Among all types of U-shaped blocks, Model O represents the one with best performance. As shown in Figure 6.6, four blocks have been identified to have the characteristics of the U-shape: Block 1, 2, 8, and 9. With the intention of applying the fourth adaptation on them, they are turned into Model O from other types of U-shaped blocks. On the one hand the results of Block 1, 2, and 8 is not obvious due to the influence of other factors, and on the other hand the impact on central part of Block 9 is highly reduced. Therefore, it can be identified as an effective and feasible meso-scale urban-form-based adaptation to ESLEs.

Overall, this research has proved that certain meso-scale urban form-based adaptations are quiet effective to reduce the impact of extreme sea level events. Meanwhile, through the application case study in this chapter it is proved that such adaptations can be integrated within waterfront development proposals.

Chapter 7 Conclusions

This chapter presents the summary and conclusion of this research. Firstly, an overview of this study is addressed through reviewing the research motivations, objectives and the adopted methodology. Secondly, the main findings of this research are summarized. Thirdly, the contributions and implications of this research are discussed. The limitations and recommendations for further research are discussed in the final section.

7.1 Research overview

With the perspective combined with waterfront development, urban sustainability, and climate change, this study examined the effectiveness of the urban form-based adaptations to extreme sea level events proposed by previous studies. Considering the rising sea level and changing climate, extreme sea level events like storm surge floods and tsunamis may increase in terms of frequency and intensity. Waterfront developments will be seriously threatened by coastal disasters. In order to respond, on one hand, coastal protections should be enhanced, which includes adding more sea walls and restoring coastal ecological system. On the other hand, other adaptation strategies should be applied on waterfront developments since protections may not be sufficiently effective.

What is the role of urban designers and planners in responding to extreme sea level events?Can adaptations based on urban forms effectively reduce the impact of extreme sea level events on waterfront developments? Can such adaptations be integrated with the urban design and physical planning of waterfronts? With such research questions proposed in this thesis, a research gap is identified in the study of effective urban form-based adaptations, especially in the meso-scale. Therefore, the main research objective is to reveal effective and feasible meso-scale urban form-based adaptations through scientific methodology. Through the literature review and previous studies, five hypotheses proposed, which are related to the aspects of urban block typologies, development intensity, block depths, block orientations, and various U-shaped blocks.

CFD simulations are employed in this research as the main method to examine the hypotheses. Maximum Force-x and maximum FXPUW are selected as the variables to compare the impact on the buildings. Through five experiments, four urban form-based adaptations are theoretically verified to be effective to reduce the impact on buildings during ESLEs except the one related to block depths. In addition, an application case study is designed based on the waterfront development proposal for Marina South, Singapore in order to further test their effectiveness and feasibility.

7.2 Main findings

The simulations indicate that ESLEs can generate great impacts on the buildings in waterfronts if the protection measures fail to prevent the waves splashing into cities. The results also show that the impact on the buildings in the front row is greater than that at the back. In other words, buildings which lie closest to seawater bear the brunt of ESLEs. It illustrates that adaptations for waterfront development should focus on reducing the impact on those buildings. According to the analyses, four meso-scale urban form-based adaptations have been demonstrated to be effective in response to ESLEs.

7.2.1 Main findings of experiments

1) Changing urban blocks' typologies
The impact of ESLEs on the buildings close to seawater can be reduced by more than 10% if the urban block typology is changed from scattered ones (such as the typologies represented by Model SP and BP) to the compact ones (such as mega blocks, courtyards and street slabs).

2) Adjusting the site coverage ratio of waterfront developments
Secondly, adjusting the site coverage ratio of waterfronts into an optimum range also contributes to enhancing their resiliency without compromising the utilization efficiency of coastal land. This research discovers that for a waterfront composed by small scaled buildings, the optimum site coverage ratio is around 36%.

3) Shifting the blocks' orientations
Thirdly, shifting the buildings' orientation is also an effective method to enhance waterfronts' capacity to respond to ESLEs. For instance, if the angle between the shoreline and facades is shifted from 0° to 30°, the impact of ESLEs can be re-

duced by 49.7% in some cases.

4) Modifying U-shaped blocks

Last but not least, this research also proves that modifying U-shaped buildings can effectively reduce the impact of ESLEs. For example, if a U-shaped building is divided into three individual parts, the impact of ESLEs can decrease by more than 10%.

However, it is difficult to prove that increasingthe depths of buildings will definitely reduce the impact of ESLEs; it is still convincing that blocks' depths are influential factors that can enhance or undermine the resiliency of waterfronts.

7.2.2 Main findings of application study

Meanwhile, the application case study asserts that meso-scale urban form based adaptations could be successfully applied in waterfront developments. Although the original urban fabric has been partly changed, it is still possible to meet the designed functions in the modified proposal. Most importantly, the comparison of the simulations' results demonstrates that the impact of ESLEs can be reduced by the application of adaptations. First of all, it shows that changing block typologies can help reduce the impact on buildings, but more impact than expected might act on the behind ones. Therefore, this adaptation should be carefully applied when the impact on the buildings in the back could not reach the threshold of causing damage. In addition, the results also show that shifting the orientation of blocks or buildings is quite effective on most of blocks. Nonetheless, there remains a concern about this adaptation since the shifting of orientation might affect the indoor and outdoor spaces of buildings. Accordingly, the rotation of orientation must be controlled within the permitted range. Besides, modifying U-shaped blockscan be identified as an effective and feasible meso-scale urban-form-based adaptation to ESLEs. Several U-shaped blocks are divided into three individual parts, on which the impact of ESLES is reduced to certain extents.

Overall, this research has proved that certain meso-scale urban form-based adaptations are quiet effective to reduce the impact of extreme sea level events. Meanwhile, through the application case study, it is proved that such adaptations can be integrated into waterfront development proposals.

7.3 Contributions and implications

7.3.1 Theoretical contributions

One key contribution of this study is that the research methodology can be adopted by other studies related to of urban form-based adaptation to ESLEs. The previously existing research regarding this topic has been carried out through a hypothetical way and the outcomes are limited to theoretical discussions. The effectiveness and feasibility of adaptations failed to be scientifically assessed. The methodology developed in this study combines typological, parametric, and case study approaches. Through CFD simulations, the impact of ESLEs on buildings can be quantitatively analyzed. Thereby, the effectiveness of adaptations could be concluded through comparing the impacts on different urban forms. This method can be widely adopted by other similar research. Most importantly, this research expands the study of waterfront development in terms of sustainable urban design. Combining the research areas of urban design, sustainable development, and climate change, the research embarks a new direction for waterfronts study. Facilitated by the methodology of this research, more potential studies regarding the performance of waterfronts' urban forms during ESLEs can be carried out.

7.3.2 Practical contribution

With the intention to enhance waterfront resiliency in calamitous climate, this study also contributes in aspects of practical implications. Nowadays, coastal cities become increasingly important to the global economic networks, and especially their waterfronts are essential parts of the cities in terms of economic activities, urban image, and residents' living quality, etc. This study contributes some knowledge for the physical planning and urban design in waterfronts in terms of physical response to extreme sea level events.

The experiment results will facilitate sustainable strategies to shape cities' waterfronts. Four meso-scale urban form-based adaptations are proved to be effective and feasible in this research. Accordingly, for the existing urban areas with vulnerable locations, these adaptations could be adopted in aspects of planning review and adjustment in order to improve thecapacity of coastal cities to deal with extreme sea level events. Moreover, for the new development schemes of coastlines or of reclaimed land, this research provides a useful assessment methodology for urban planning authorities to identify effective urban designsfor waterfront developments as well as providing modification options for the ineffective designs.

In addition, this research examines Singapore's response strategies to sea level rise and related extreme events through the case study of Marina South. Consequently, the waterfronts' capacities to deal with climate change are discussed andthe effective adaptations are integrated into Singapore's waterfront developments so that the research could provide significant design and assessment strategies to local urban planning authorities. It could be adapted by URA, BCA and some other authorities and provide complements of current regulations for waterfront developments.

7.4 Limitations and recommendations for further research

Being an exploratory study, this work has its limitations. One issue is that this study does not consider the effect of drainage system in simulations. The drainage system might be quite largely mentioned when the steady floods are caused by heavy rainfall. However, it may not play an important role when the flood waves have quite high velocity. Therefore, the omitting of drainage system in simulations might not affect the results in this study.

In addition, another limitation is that the vegetations were not included in simulation in case studies. Actually, it has been well demonstrated that vegetations, such as grass, trees and bushes, can effectively reduce the velocity of storm surge floods. For the purpose to compare the effectiveness of purely different urban forms, this study excludes the vegetations, which opens another possible research avenue to combine the effects of urban design adaptations and vegetations in the future studies.

Besides, in order to facilitate the research, only the maximum Force-x and maximum FXPUW are chosen as the variables to carry out comparisons and analyses. Although they largely represent the impact of ESLEs on buildings, other factors such as average Force-x and average FXPUW are also useful to assess the damages caused by ESLEs. Thereby, in future study, more comprehensive and multi-variables comparisons and analyses should be conducted.

Because this thesis focuses on the meso-scale adaptations, the buildings in all models are simplified as cubes. The effect of changing the buildings into other shapes is not included in this research. It is believed that cylindrical buildings might be more effective in terms of reducing the impact of ESLEs. Therefore, there is a gap for future study to reveal the different capacities of various building shapes in response to ESLEs.

With a location protected by periphery islands, Singapore's situation is better than other cities in Asian Pacific region. Some areas, especially the estuary areas are more vulnerable than Singapore. For example, after assessing a 72-year tidal record of Hong Kong and factors such as estuarine backwater effects and long-term geological subsidence, it is suggested that a 30 cm rise in relative sea level at the mouth of the estuary is possible by 2030(Z Huang et al., 2004). The increase of sea level around Pearl River Delta would destroy most of the reclaimed land of Hong Kong, Macau and Shenzhen. Therefore, with the methodology developed in this research, the future study could be carried out based on other coastal cities with high vulnerability in terms of sea level rise and storm surge floods.

7.5 Conclusion

For centuries, urbanists like Ebenezer Howard(2009), Le Corbusier(1987) and Frank Lloyd Wright(2008) have been trying to propose perfect urban models, including the garden city, the broadacre city, and the radiant city. All efforts were focused on social and environmental issues of their own times. In our era, the challenge from climate change is more exigent than ever before. As contemporary urban researchers, our responsibility shifts to develop urban models which are adaptive to climate change, and the fundamental criterion for these models is sustainability. Although this preliminary research is far away from proposing perfect models for waterfronts, it indicates that spatial arrangement tools and specific urban forms may enhance the resiliency of waterfronts. It could be considered as the initiative to develop transformational adaptations and sustainable models for coastal cities which are struggling with the rising sea.

References

Ali, A. (1996). Vulnerability of Bangladesh to climate change and sea level rise through tropical cyclones and storm surges Climate Change Vulnerability and Adaptation in Asia and the Pacific (pp. 171-179): Springer.

Amos, L. W. (2009). Unravelling thresholds: Mixed use developments for downtown waterfronts. Dalhousie University.

Anderson, J. B. (2007). The formation and future of the upper Texas coast : a geologist answers questions about sand, storms, and living by the sea (1st ed.). College Station, Tex.: Texas A & M University Press.

Anderson, J. D., & Wendt, J. F. (1995). Computational fluid dynamics (Vol. 206): McGraw-Hill.

Ballentine, T. B. (2006). A Look at the Past, Present, and Future Roles of Historic Preservation in Seattle's Central Waterfront. University of Washington.

Berkes, F. (2006). From community-based resource management to complex systems: the scale issue and marine commons. Ecology and Society, 11(1), 45.

Biesbroek, G. R., Swart, R. J., Carter, T. R., Cowan, C., Henrichs, T., Mela, H., et al. (2010). Europe adapts to climate change: comparing national adaptation strategies. Global Environmental Change, 20(3), 440-450.

Blowers, A. (2013). Planning for a sustainable environment: Routledge.

Bone, K., Betts, M. B., & Greenberg, S. (1997). The New York waterfront: evolution and building culture of the port and harbor: Monacelli Pr.

Breen, A., & Rigby, D. (1996). The new waterfront: a worldwide urban success story: McGraw-Hill Professional.

Breen, A., Rigby, D., Norris, D. C., & Norris, C. (1994). Waterfronts: Cities reclaim their edge: McGraw-Hill New York.

Brundtland, G. H. (1987). World commission on environment and development. Our common future, 8-9.

Bush, D. M., Pilkey, O. H., & Neal, W. J. (1996). Living by the rules of the sea. Durham, N.C.: Duke University Press.

Bush, D. M., Pilkey, O. H., & Neal, W. J. (1996). Living by the Rules of the Sea: Duke University Press Books.

Cai, F., Su, X., Liu, J., Li, B., & Lei, G. (2009). Coastal erosion in China under the condition of global climate change and measures for its prevention. Progress in Natural Science, 19(4), 415-426.

Caleffi, V., Valiani, A., & Zanni, A. (2003). Finite volume method for simulating extreme flood events in natural channels. Journal of Hydraulic Research, 41(2), 167-177.

Calthorpe, P., & Fulton, W. B. (2001). The regional city: Planning for the end of sprawl: Island Pr.

Carter, A. (2000). Strategy and partnership in urban regeneration. Urban regeneration: a handbook, 37-58.

Carter, H. (1986). 10 Cardiff: local, regional and national capital. Regional cities in the UK, 1890-1980, 171.

Cash, D. W., Adger, W. N., Berkes, F., Garden, P., Lebel, L., Olsson, P., et al. (2006). Scale and cross-scale dynamics: governance and information in a multi-level world. Ecology and Society, 11(2), 8.

Chan, Y. M. (1999). The impacts of sea level rise on the coasts of Singapore. National University of Singapore, Singapore.

Charlesworth, J., & Cochrane, A. (1994). Tales of the suburbs: the local politics of growth in the South-East of England. Urban Studies, 31(10), 1723.

Chen, J. (1997). The Impact of Sea Level Rise on China's Coastal Areas and Its Disaster Hazard Evaluation. Journal of Coastal Research, 13(3), 925-930.

Church, J. A., Hunter, J. R., McInnes, K. L., & White, N. J. (2006). Sea-level rise around the Australian coastline and the changing frequency of extreme sea-level events. Australian Meteorological Magazine, 55(4), 253-260.

Cicin-Sain, B. (1993). Sustainable development and integrated coastal management. Ocean & coastal management, 21(1-3), 11-43.

Cohen, J., Small, C., Mellinger, A., Gallup, J., Sachs, J., Vitousek, P., et al. (1997). Estimates of coastal populations. Science, 278, 1209.

Cohen, N. (2011). Green Cities: An A-to-Z Guide (Vol. 4): Sage.

Commission, D. (2008). Working Together with Water. A Living Land Builds for its Future. Findings of the Deltacommissie.

Corbusier, L. (1987). The city of to-morrow and its planning: Courier Dover Publications.

Couch, C. (1990). Urban renewal: theory and practice: Macmillan Education.

Craig-Smith, S. J., & Fagence, M. (1995). Recreation and tourism as a catalyst for urban waterfront redevelopment: an international survey: Praeger Publishers.

Davoudi, S., Crawford, J., & Mehmood, A. (2009). Planning for climate change : strategies for mitigation and adaptation for spatial planners. London ; Sterling, VA: Earthscan.

Davoudi, S., Crawford, J., & Mehmood, A. (2009). Planning for climate change: strategies for mitigation and adaptation for spatial planners: Earthscan/James & James.

Dean, R. G. (2002). Beach nourishment: theory and practice (Vol. 18): World Scientific.

Dessai, S., & Hulme, M. (2004). Does climate adaptation policy need probabilities? Climate Policy, 4(2), 107–128.

Dossou, K. M., & Glehouenou-Dossou, B. (2007). The vulnerability to climate change of Cotonou (Benin) the rise in sea level. Environment and Urbanization, 19(1), 65–79.

Doucet, B. (2010). Rich cities with poor people: waterfront regeneration in the Netherlands and Scotland. Netherlands Geographical Studies, 391.

Dronkers, J., Gilbert, J., Butler, L., Carey, J., Campbell, J., James, E., et al. (1990). Strategies for Adaptation to Sea Level Rise.

Dwarakish, G. S., Vinay, S. A., Natesan, U., Asano, T., Kakinuma, T., Venkataramana, K., et al. (2009). Coastal vulnerability assessment of the future sea level rise in Udupi coastal zone of Karnataka state, west coast of India. Ocean & coastal management, 52(9), 467–478.

Edwards, J. A. (1996). Waterfronts, tourism and economic sustainability: the United Kingdom experience. Sustainable Tourism, 86–98.

Environment-Agency. (2009). Thames Estuary 2100: Managing Flood Risk through London and the Thames Estuary.

ESC. (2010). ECONOMIC STRATEGIES COMMITTEE KEY RECOMMENDATIONS. Singapore Economic Strategies Committee.

Fan, S., Gloor, M., Mahlman, J., Pacala, S., Sarmiento, J., Takahashi, T., et al. (1998). A large terrestrial carbon sink in North America implied by atmospheric and oceanic carbon dioxide data and models. Science, 282(5388), 442–446.

Field, C. B., Barros, V., & Stocker, T. F. (2012). Managing the risks of extreme events disasters to advance climate change adaptation. Recherche, 67, 02.

Geisser, S., & Johnson, W. O. (2006). Modes of parametric statistical inference (Vol. 529): John Wiley & Sons.

Gosling, D., & Gosling, M. C. (2003). The evolution of American urban design: a chronological anthology: Academy Press.

Gospodini, A. (2001). Urban Waterfront Redevelopment in Greek Cities:: A Framework for Redesigning Space. Cities, 18(5), 285-295.

Grammenos, F., Pogharian, S., & Tasker-Brown, J. (2002). Residential street pattern design. Socio-economic Series, 75, 22.

Greenberg, K. (1996). WATERFRONT AS A TERRAIN OF AVAILABILITY. City, capital, and water, 195.

Gupta, V. (1984). Solar radiation and urban design for hot climates. Environment and Planning B: Planning and Design, 11(4), 435-454.

Gupta, V. (1987). Thermal efficiency of building clusters: an index for non air-conditioned buildings in hot climates. Energy and urban built form, 133.

Haider, S., Paquier, A., Morel, R., & Champagne, J. (2003). Urban flood modelling using computational fluid dynamics.

Hall, P. (1991). Waterfronts: a new urban frontier.

Hamilton, W. G., & Simard, B. (1993). Victoria's Inner Harbour 1967©\1992: The Transformation Of A Deindustrialized Waterfront. Canadian Geographer/Le G¨ ¡ographe canadien, 37(4), 365-371.

Har, L. I. S. S. (1997). CONSERVATION ISSUES IN WATERFRONT DEVELOPMENT.

Harvey, D. (1974). Class-monopoly rent, finance capital and the urban revolution. Regional Studies: The Journal of the Regional Studies Association, 8, 3(4), 239-255.

Heberger, M., Cooley, H., Herrera, P., Gleick, P. H., & Moore, E. (2009). The impacts of sea-level rise on the California coast: Pacific Institute.

Heberger, M., Cooley, H., Herrera, P., Gleick, P. H., & Moore, E. (2011). Potential impacts of increased coastal flooding in California due to sea-level rise. Climatic Change, 109(1), 229-249.

Hines, T. S., & Harris, N. (2008). Burnham of Chicago: architect and planner: University of Chicago Press.

Ho, J. (2008). Singapore Country Report—A Regional Review on the Economics of Climate Change in Southeast Asia. Report submitted for RETA, 6427.

Hoyle, B. (1995). A Shared Space: Contrasted Perspectives on Urban Waterfront Redevelopment in Canada. The Town Planning Review, 66(4), 345-369.

Hoyle, B. (2000). Global and Local Change on the Port-City Waterfront. Geographical Review, 90(3), 395-417.

Hoyle, B. S., Pinder, D., & Husain, M. S. (1988). Revitalising the waterfront: international dimensions of dockland redevelopment: Belhaven Press.

Hu., W. N. (2004). Study of Spatial Form of Two-bank-waterfront City. Tsinghua University

Huang, Z., Wu, T. R., Tan, S. K., Megawati, K., Shaw, F., Liu, X., et al. (2009). Tsunami hazard from the subduction Megathrust of the South China Sea: Part II. Hydrodynamic modeling and possible impact on Singapore. Journal of Asian Earth Sciences, 36(1), 93-97.

Huang, Z., Zong, Y., & Zhang, W. (2004). Coastal inundation due to sea level rise in the Pearl River Delta, China. Natural Hazards, 33(2), 247-264.

Huq, S., Kovats, S., Reid, H., & Satterthwaite, D. (2007). Editorial: Reducing risks to cities from disasters and climate change. Environment and Urbanization, 19(1), 3-15.

Hurlimann, A., Barnett, J., Fincher, R., Osbaldiston, N., Mortreux, C., & Graham, S. (2014). Urban planning and sustainable adaptation to sea-level rise. Landscape and urban planning.

IPCC. (2007). Climate Change 2007: Synthesis ReportSummary for Policymakers. Jutla, R. S. (2000). Visual image of the city: tourists' versus residents' perception of Simla, a hill station in northern India. Tourism Geographies, 2(4), 404-420.

Kasintiz, P., & Rosenberg, J. (1996). Missing the connection: Social isolation and employment on the Brooklyn waterfront. Soc. Probs., 43, 180.

Kates, R. W., Travis, W. R., & Wilbanks, T. J. (2012). Transformational adaptation when incremental adaptations to climate change are insufficient. Proceedings of the National Academy of Sciences, 109(19), 7156-7161.

Kelletat, D. (1992). Coastal erosion and protection measures at the German North Sea coast. Journal of Coastal Research, 699-711.

Kemp, R. L. (2003). Community renewal through municipal investment: a handbook for citizens and public officials: McFarland.

Kolomiets, P., Zheleznyak, M., & Tkalich, P. (2010). Dynamical Downscaling of Storm Surges in South-China Sea and Singapore Strait. Paper presented at the Storm Surges Congress.

LIM., S. L. (2005). Waterfront Revitalization in Singapore: Transformation of Public Space and Creation of Landscape Spectacles National Universtiy of Singapre, Singapore.

Lin-xue, L. (1999). Evolution of Spatial Morphology of Urban Waterfront [J]. TIME+ARCHITECTURE, 3.

Lorenzoni, I., Nicholson-Cole, S., & Whitmarsh, L. (2007). Barriers perceived to engaging with climate change among the UK public and their policy implications. Global Environmental Change, 17(3), 445-459.

Lynas, M. (2008). Six degrees: Our future on a hotter planet: National Geographic

Books.

Lynch, K. (1992). The image of the city: MIT press.

Lynch, K., & Rodwin, L. (1958). A theory of urban form. Journal of the American Institute of Planners, 24(4), 201-214.

Ma, L., Ashworth, P. J., Best, J. L., Elliott, L., Ingham, D. B., & Whitcombe, L. J. (2002). Computational fluid dynamics and the physical modelling of an upland urban river. Geomorphology, 44(3-4), 375-391.

March, L., & Trace, M. (1968). The land use performances of selected arrays of built forms: University of Cambridge,[Centre for Land Use and Built Form Studies. Marcus, C. C., & Francis, C. (1997). People places: Design guidelines for urban open space: Wiley.

Marshall, R. (2001a). Contemporary urban space-making at the water's edge. Waterfronts in Post-Industrial Cities, 3-14.

Marshall, R. (2001b). Waterfronts in post-industrial cities: Taylor & Francis.
Martin, L., & March, L. (1972). Urban space and structures (Vol. 1): University Press.

McCarthy, J. (1998). Waterfront regeneration: recent practice in Dundee. European Planning Studies, 6(6), 731-736.

McCarthy, J. J. (2001). Climate change 2001: impacts, adaptation, and vulnerability: contribution of Working Group II to the third assessment report of the Intergovernmental Panel on Climate Change: Cambridge Univ Pr.

Mei-e, R. (1993). Relative sea-level changes in China over the last 80 years. Journal of Coastal Research, 229-241.

Meyer, H. (1999a). City and port: International Books.

Meyer, H. (1999b). City and port—transformation of port cities: London, Barcelona, New York, Rotterdam. Utrecht: International Books.

Millspaugh, M. L. (2001). Waterfronts as catalysts for city renewal. Waterfronts in Post-Industrial Cities, 74-85.

Minca, C. (1995). Urban Waterfront Evolution: The Case of Trieste. Geography, 80(3), 225-234.

Monaghan, J. J. (1994). Simulating free surface flows with SPH. Journal of computational physics, 110, 399-399.

Montavon, M., Steemers, K., Cheng, V., & Compagnon, R. (2006). La Ville Radieuse'by Le Corbusier: Once again a case study. PLEA, Geneva, Switzerland.

NCCS. (2008). Singapore's National Climate Change Strategy.

NCCS. (2012). Climate Change and Singapore: Challenges. Opportunities. Partnerships. Singapore: National Climate Change Secretariat

Neil Adger, W., Arnell, N. W., & Tompkins, E. L. (2005). Successful adaptation to climate change across scales. Global Environmental Change, 15(2), 77-86.

Ng, E. (2005). A study of the relationship between daylight performance and height difference of buildings in high density cities using computational simulation. Paper presented at the International Building Performance Simulation Conference.

Ng, W. S., & Mendelsohn, R. (2005). The impact of sea level rise on Singapore. Environment and Development Economics, 10(2), 201-215.

Ng, W. S., & Mendelsohn, R. (2006). The economic impact of sea-level rise on nonmarket lands in Singapore. Ambio, 35(6), 289-296.

Nicholls, R., Hanson, S., Herweijer, C., Patmore, N., Hallegatte, S., Corfee-Morlot, J., et al. (2008). Ranking port cities with high exposure and vulnerability to climate extremes. Organization for Economic Development, 19.

Nicholls, R., & Mimura, N. (1998). Regional issues raised by sea-level rise and their policy implications. Climate Research, 11, 5-18.

Nicholls, R. J., Hoozemans, F. M., & Marchand, M. (1999). Increasing flood risk and wetland losses due to global sea-level rise: regional and global analyses. Global Environmental Change, 9, S69-S87.

NPB. (2009). National Biodiversity Strategy and Action Plan

Nurse, L. A., Sem, G., Hay, J., Suarez, A., Wong, P. P., Briguglio, L., et al. (2001). Small island states (pp. 843-876): Cambridge University Press, Books, 40 West 20 th Street New York NY 10011-4211 USA.

Osbahr, H., Twyman, C., Neil Adger, W., & Thomas, D. S. (2008). Effective livelihood adaptation to climate change disturbance: scale dimensions of practice in Mozambique. Geoforum, 39(6), 1951-1964.

Pacala, S. W., Hurtt, G. C., Baker, D., Peylin, P., Houghton, R. A., Birdsey, R. A., et al. (2001). Consistent land-and atmosphere-based US carbon sink estimates. Science, 292(5525), 2316-2320.

Pan, Y., Birdsey, R. A., Fang, J., Houghton, R., Kauppi, P. E., Kurz, W. A., et al. (2011). A large and persistent carbon sink in the world's forests. Science, 333(6045), 988-993.

Paris, R., Lavigne, F., Wassmer, P., & Sartohadi, J. (2007). Coastal sedimentation associated with the December 26, 2004 tsunami in Lhok Nga, west Banda Aceh (Sumatra, Indonesia). Marine Geology, 238(1-4), 93-106.

Park, H., Cox, D. T., Lynett, P. J., Wiebe, D. M., & Shin, S. (2013). Tsunami inundation modeling in constructed environments: A physical and numerical comparison of free-surface elevation, velocity, and momentum flux. Coastal Engineering, 79, 9-21.

Parry, M. L. (2007). Climate Change 2007: Impacts, Adaptation and Vulnerability: Working Group II Contribution to the Fourth Assessment Report of the IPCC Intergovernmental Panel on Climate Change (Vol. 4): Cambridge University Press.

Pender, G., Morvan, H., Wright, N., & Ervine, D. (2005). CFD for environmental design and management. Computational Fluid Dynamics, 487-509.

Plachouras, V., & Ounis, I. (2007). Longman dictionary of contemporary English. Paper presented at the in 'Proceedings of the 29th European Conference on Information Retrieval (ECIR' 07).

Qing, K. G. (2005). Singapore Finds it Hard to Expand Without Sand. Wild Singapore, 050412-050411.

Rahmstorf, S. (2007). A semi-empirical approach to projecting future sea-level rise. Science, 315(5810), 368-370.

Ratti, C., Raydan, D., & Steemers, K. (2003). Building form and environmental performance: archetypes, analysis and an arid climate. Energy and Buildings, 35(1), 49-59.

Ratti, C., & Richens, P. (2004). Raster analysis of urban form. Environment and Planning B, 31(2), 297-310.

Reid, H., & Huq, S. (2007). How we are set to cope with the impacts. Adaptation to climate change. IIED Briefing, International Institute of Environment and Development, London.

Risselada, M., & van den Heuvel, D. (2005). Team 10: 1953-81, in Search of a Utopia of the Present: NAi.

Roberts, P. (2000). The evolution, definition and purpose of urban regeneration. Urban regeneration: a handbook, 9-36.

Rose-redwood, R. S. (2008). Genealogies of the grid: revisiting stanislawski's search for the origin of the grid-pattern town*. Geographical Review, 98(1), 42-58.

Ruskeepää, L. A. D. (2011). Adaptation and adaptability: expectant design for resilience in coastal urbanity. Massachusetts Institute of Technology.

Sampei, Y., & Aoyagi-Usui, M. (2009). Mass-media coverage, its influence on public awareness of climate-change issues, and implications for Japan's national campaign to reduce greenhouse gas emissions. Global Environmental Change, 19(2), 203-212.

Schiffman, H. S. (2011). Green Issues and Debates: An A-to-Z Guide (Vol. 12): SAGE.

Seguchi, T., & Malone, P. (1996). Tokyo: waterfront development and social needs. City, Capital and Water, London, Routledge, 164-194.

Serageldin, M. (1997). A decent life. Harvard Design Magazine, 40-41.

Sharma, D., & Tomar, S. (2010). Mainstreaming climate change adaptation in Indian cities. Environment and Urbanization, 22(2), 451-465.

Smith, N., & Williams, P. (1986). Gentrification of the City: Allen & Unwin London.

Sofield, T. H. B., & Sivan, A. (1994). From cultural festival to international sport-the Hong Kong Dragon Boat Races. Journal of Sport Tourism, 1(3), 5-17.

Stanback, T. M., & Knight, R. V. (1976). SUBURBANIZATION AND THE CITY: Allanheld, Osmun; distribution, Universe Books (Montclair, NJ).

Steemers, K., Baker, N., Crowther, D., Dubiel, J., Nikolopoulou, M., & Ratti, C. (1997). City texture and microclimate. Urban Design Studies, 3, 25-50.

Stocker, T., Qin, D., Plattner, G., Tignor, M., Allen, S., Boschung, J., et al. (2013). IPCC, 2013: Summary for Policymakers. In: Climate Change 2013: The Physical Science Basis. Contribution of Working Group I to the Fifth Assessment Report of the Intergovernmental Panel on Climate Change: Cambridge Univ Press, Cambridge, United Kingdom and New York, NY, USA.

Tice, J. (1993). Theme and Variations: A Typological Approach to Housing Design, Teaching, and Research. Journal of architectural education, 46(3), 162-175.

Tkalich, P., Vethamony, P., Babu, M., & Pokratath, P. (2009). Seasonal sea level variability and anomalies in the Singapore Strait.

Torre, L. A. (1989). Waterfront development: Van Nostrand Reinhold.

Tweedale, I. (1988). Waterfront redevelopment, economic restructuring and social impact. Revitalizing the Waterfront, 185-198.

URA. (2001). The Concept Plan 2001: Urban Redevelopment Authority.

URA. (2007). Waterfront-Garden living planned at Marina South. from http://www.ura.gov.sg/pr/text/2007/pr07-97.html

URA. (2010). Singapore 2011Concept Plan. Singapore.

Usavagovitwong, N., & Posriprasert, P. (2006). Urban poor housing development on Bangkok's waterfront: securing tenure, supporting community processes. Environment and Urbanization, 18(2), 523-536.

Vallega, A. (2001). Urban waterfront facing integrated coastal management. Ocean & coastal management, 44(5), 379-410.

Vasey-Ellis, N. (2009). Planning for climate change in coastal Victoria. Urban Policy and Research, 27(2), 157-169.

Viessman, W., Lewis, G. L., & Knapp, J. W. (1977). Introduction to hydrology: IEP-Dun-Donnelley, Harper & Row.

Walsh, K., Betts, H., Church, J., Pittock, A., McInnes, K., Jackett, D., et al. (2004). Using sea level rise projections for urban planning in Australia. Journal of Coastal Research, 586-598.

Walsh, K., & Pittock, A. (1998). Potential changes in tropical storms, hurricanes, and extreme rainfall events as a result of climate change. Climatic Change, 39(2), 199-213.

Walsh, K. J. E., Betts, H., Church, J., Pittock, A. B., McLnnes, K. L., Jackett, D. R., et al. (2004). Using Sea Level Rise Projections for Urban Planning in Australia. Journal of Coastal Research, 20(2), 586-598.

White, K. (1993). Urban waterside regeneration: problems and prospects: E. Horwood.

Wilbanks, T., Yohe, G., Mengelt, C., & Casola, J. (2010). America's Climate Choices: Adapting to the Impacts of Climate Change. Paper presented at the AGU Fall Meeting Abstracts.

Wilby, R. L., & Dessai, S. (2010). Robust adaptation to climate change. Weather, 65(7), 180-185.

Wilson, E., & Piper, J. (2010). Spatial Planning and Climate Change: Taylor & Francis.

Wong, P. P. (1992). Impact of a sea level rise on the coasts of Singapore: preliminary observations. Journal of Southeast Asian Earth Sciences, 7(1), 65-70.

Wrenn, D. M. (1983). Urban waterfront development. . Mary's LJ, 15, 555.

Wrenn, D. M., Casazza, J., & Smart, E. (1983). Urban waterfront development: Urban Land Inst.

Wu, Q. 中国古代城市防洪研究[M].中国建筑工业出版社，1995.

Yalciner, A. C., Ozer, C., Zaytsev, A., Suppasri, A., Mas, E., Kalligeris, N., et al. (2011). FIELD SURVEY ON THE COASTAL IMPACTS OF MARCH 11, 2011 GREAT EAST JAPAN TSUNAMI.

Yang, D. (2006). Waterfronts: Spatial composition and cultural use. Unpublished PhD thesis, University College London.

Yim, S. C. (2005). Modeling and simulation of tsunami and storm surge hydrodynamic loads on coastal bridge structures.

Yong, K., Lee, S., & Karunaratne, G. (1991). Coastal reclamation in Singapore: a review.

Zhang, J., Heng, C. K., Malone-Lee, L. C., Hii, D. J. C., Janssen, P., Leung, K. S., et al. (2012). Evaluating environmental implications of density: A comparative case study on the relationship between density, urban block typology and sky exposure. Automation in construction, 22, 90-101.

Zhuang, J. (2010). The Economics of Climate Change in Southeast Asia: A Regional Review.

Appendixes

Appendix A: Simulation settings for Experiment 1

The following information is the simulations setting for Experiment 1 (take Model SP as an example).

* Analysis Types
  - Incompressible/Compressible flow   Incompressible
  - Flow field   Turbulent flow
  - Turbulence model   Standard k-eps model
  - Flow   Consider
  - Free surface   MARS method
  - Steady Analysis/Transient Analysis   Transient analysis
* Basic Settings
  - Gravity   Consider
  - Acceleration due to gravity   (0, 0, -1) 9.8 [m/s^2]
* Fluid Region
  - Fluid number 1   Domain(cuboid) : air(20C)
  - fluid number 2 (VOF2)   water(20C)
* Flow
  - Components of velocity   X, Y, Z-direction
* Initial Condition
  - Undefined:Initial temperature of solid parts   20 C
  - The number of initial condition settings   1
  - water_vol   Initial1 : Initial VOF2
* Flow Boundary
  - The number of flux boundary settings   2
  - Xmax   Flux2 : Natural outflow boundary
  - Zmax   Flux3 : Static pressure boundary
* Wall Boundary
  - The number of wall boundary settings   3
  - Ymin   Wall3 :Freeslip
  - Ymax   Wall3 :Freeslip
  - Undefined(Stress: All fluid boundary)   Wall2 : Noslip(smooth)

* Symmetrical Boundary
    The number of symmetry boundary settings   None
* Source Condition
    The number of volumetric source condition settings   None
    The number of area source condition settings   None
* Fixed Condition
    The number of fixed velocity condition settings   None
* Free Surface
    Contact angle at wall (entire domain)   90 degrees
    Fractional steps   5 times
    Cut-off value of VOF   0.0001
    Values of VOF under the specified cut-off value   Save
    Control Courant Number in transient analysis   Minimum=0.1 Maximum=0.5
    The number of contact angle condition settings   None
    The number of initial and fixed condition settings   1
    water_vol   Initial1 : Initial VOF2
    The number of permeable body and attenuation zone condition settings   1
    attenuation-zone   FreeSurface5 : Free surface: porous media
    The number of wave generation condition settings   1
    water   wave : Free surface: wave generation
    Output Surface geometry   Output
    Output format   Time series CSV format
    Direction to search free surface   X-direction
* Analysis Control
    Type of setting   Detailed setting
* Transient Analysis
    Cycle   Initial calc.: cycles 1 to 15000
    Time step   Automatically calculated: initial time step=0.0001 Courant no.=0.3
    Specified time   360 sec
    The number of stop point settings   None
* Solver Parameters
    Matrix solver/Advection term
    X-component of velocity   Default 1st order upwind
    Y-component of velocity   Default 1st order upwind
    Z-component of velocity   Default 1st order upwind
    Pressure   Default ---

        Turbulent kinetic energy   Default 1st order upwind
        Turbulent dissipation rate   Default 1st order upwind
* Field File
        Analysis variable output   Default
* Time Series
        The number of time series settings   None
* List File Output
        Output sum of pressure   1 cycles
        The number of output settings of the sum   16
        Output region   f1
        Output region   f2
        Output region   f3
        Output region   f4
        Output region   f2-1
        Output region   f2-2
        Output region   f2-3
        Output region   f2-4
        Output region   f3-1
        Output region   f3-2
        Output region   f3-3
        Output region   f3-4
        Output region   f4-1
        Output region   f4-2
        Output region   f4-3
        Output region   f4-4
        Warning/error message   Output
        Cycle information   Output every cycles
        Matrix calculation information   Output every 1 cycles
* Output Files
        Project name   small_blocks
        Generic name for Postprocessor file   small_blocks
        Restart file (output)   small_blocks.r
        Free surface location file   small_blocks_sufl_tm.csv
        Comments   project no.1
        Field data output cycle   Constant interval: 0.5 seconds + last cycle
        Initial value for field file   Output
        Field file output format   Default(FLD file)

Restart output cycle　100 cycles

Appendix B: Simulation settings for Experiment 2

The following information is the simulations setting for Experiment 1 (take Model A as an example).

* Analysis Types
　　　Incompressible/Compressible flow　　Incompressible
　　　Flow field　　Turbulent flow
　　　Turbulence model　　　Standard k-eps model
　　　Flow　Consider
　　　Free surface　MARS method
　　　Steady Analysis/Transient Analysis　　Transient analysis
* Basic Settings
　　　Gravity　Consider
　　　Acceleration due to gravity　　(0, 0, -1) 9.8 [m/s2]
* Fluid Region
　　　Fluid number 1　　Domain(cuboid) : air(20C)
　　　fluid number 2 (VOF2)　　water(20C)
* Flow
　　　Components of velocity　　X, Y, Z-direction
* Initial Condition
　　　Undefined:Initial temperature of solid parts　　20 C
　　　The number of initial condition settings　　1
　　　water_vol　Initial1 : Initial VOF2
* Flow Boundary
　　　The number of flux boundary settings　　2
　　　Xmax　Flux2 : Natural outflow boundary
　　　Zmax　Flux3 : Static pressure boundary
* Wall Boundary
　　　The number of wall boundary settings　　3
　　　Ymin　Wall3 :Freeslip
　　　Ymax　Wall3 :Freeslip
　　　Undefined(Stress: All fluid boundary)　　Wall2 : Noslip(smooth)
* Symmetrical Boundary
　　　The number of symmetry boundary settings　　None
* Source Condition

　　　　The number of volumetric source condition settings　None
　　　　The number of area source condition settings　None
* Fixed Condition
　　　　The number of fixed velocity condition settings　None
* Free Surface
　　　　Contact angle at wall (entire domain)　90 degrees
　　　　Fractional steps　5 times
　　　　Cut-off value of VOF　0.0001
　　　　Values of VOF under the specified cut-off value　Save
　　　　Control Courant Number in transient analysis　Minimum=0.1 Maximum=0.5
　　　　The number of contact angle condition settings　None
　　　　The number of initial and fixed condition settings　1
　　　　water_vol　Initial1 : Initial VOF2
　　　　The number of permeable body and attenuation zone condition settings　1
　　　　attenuation-zone　FreeSurface5 : Free surface: porous media
　　　　The number of wave generation condition settings　1
　　　　water　wave : Free surface: wave generation
　　　　Output Surface geometry　Output
　　　　Output format　Time series CSV format
　　　　Direction to search free surface　X-direction
* Analysis Control
　　　　Type of setting　Detailed setting
* Transient Analysis
　　　　Cycle　Initial calc.: cycles 1 to 5000
　　　　Time step　Automatically calculated: initial time step=0.0001 Courant no.=0.3
　　　　Specified time　360 sec
　　　　The number of stop point settings　None
* Solver Parameters
　　　　Matrix solver/Advection term
　　　　X-component of velocity　Default 1st order upwind
　　　　Y-component of velocity　Default 1st order upwind
　　　　Z-component of velocity　Default 1st order upwind
　　　　Pressure　Default ---
　　　　Turbulent kinetic energy　Default 1st order upwind
　　　　Turbulent dissipation rate　Default 1st order upwind
* Field File

            Analysis variable output   Default
* Time Series
            The number of time series settings   None
* List File Output
            Output sum of pressure   1 cycles
            The number of output settings of the sum   4
            Output region   f1
            Output region   f2
            Output region   f2-1
            Output region   f2-2
            Warning/error message   Output
            Cycle information   Output every cycles
            Matrix calculation information   Output every 1 cycles
* Output Files
            Project name   16-1_v11
            Generic name for Postprocessor file   16-1_v11
            Restart file (output)   16-1_v11.r
            Free surface location file   16-1_v11_sufl_tm.csv
            Comments   project no.1
            Field data output cycle   Constanct interval: 0.5 seconds + last cycle
            Initial value for field file   Output
            Field file output format   Default(FLD file)
            Restart output cycle   100 cycles

Appendix C: Simulation settings for Experiment 3

The following information is the simulations setting for Experiment 1 (take Model G as an example).
* Analysis Types
            Incompressible/Compressible flow   Incompressible
            Flow field   Turbulent flow
            Turbulence model   Standard k-eps model
            Flow   Consider
            Free surface   MARS method
            Steady Analysis/Transient Analysis   Transient analysis
* Basic Settings
            Gravity   Consider

Acceleration due to gravity (0, 0, -1) 9.8 [m/s^2]
* Fluid Region
    Fluid number 1   Domain(cuboid) : air(20C)
    fluid number 2 (VOF2)   water(20C)
* Flow
    Components of velocity   X, Y, Z-direction
* Initial Condition
    Undefined:Initial temperature of solid parts   20 C
    The number of initial condition settings   1
    water_vol   Initial1 : Initial VOF2
* Flow Boundary
    The number of flux boundary settings   2
    Xmax   Flux2 : Natural outflow boundary
    Zmax   Flux3 : Static pressure boundary
* Wall Boundary
    The number of wall boundary settings   3
    Ymin   Wall4 :Freeslip
    Ymax   Wall4 :Freeslip
    Undefined(Stress: All fluid boundary)   Wall2 : Noslip(smooth)
* Symmetrical Boundary
    The number of symmetry boundary settings   None
* Source Condition
    The number of volumetric source condition settings   None
    The number of area source condition settings   None
* Fixed Condition
    The number of fixed velocity condition settings   None
* Free Surface
    Contact angle at wall (entire domain)   90 degrees
    Fractional steps   5 times
    Cut-off value of VOF   0.0001
    Values of VOF under the specified cut-off value   Save
    Control Courant Number in transient analysis   Minimum=0.1 Maximum=0.5
    The number of contact angle condition settings   None
    The number of initial and fixed condition settings   1
    water_vol   Initial1 : Initial VOF2
    The number of permeable body and attenuation zone condition settings   1
    attenuation-zone   FreeSurface5 : Free surface: porous media

The number of wave generation condition settings   1
water   wave : Free surface: wave generation
Output Surface geometry   Output
Output format   Time series CSV format
Direction to search free surface   X-direction

* Analysis Control
    Type of setting   Detailed setting
* Transient Analysis
    Cycle   Initial calc.: cycles 1 to 20000
    Time step   Automatically calculated: initial time step=0.0001 Courant no.=0.3
    Specified time   360 sec
    The number of stop point settings   None
* Solver Parameters
    Matrix solver/Advection term
    X-component of velocity   Default 1st order upwind
    Y-component of velocity   Default 1st order upwind
    Z-component of velocity   Default 1st order upwind
    Pressure   Default ---
    Turbulent kinetic energy   Default 1st order upwind
    Turbulent dissipation rate   Default 1st order upwind
* Field File
    Analysis variable output   Default
* Time Series
    The number of time series settings   None
* List File Output
    Output sum of pressure   1 cycles
    The number of output settings of the sum   17
    Output region   water
    Output region   f1-1
    Output region   f1-2
    Output region   f2-1
    Output region   f2-2
    Output region   f1-3
    Output region   f1-4
    Output region   f2-3
    Output region   f2-4
    Warning/error message   Output

   Cycle information Output every cycles
   Matrix calculation information Output every 1 cycles
* Output Files
   Project name test1021-25-35_v11
   Generic name for Postprocessor file test1021-25-35_v11
   Restart file (output) test1021-25-35_v11.r
   Free surface location file test1021-25-35_v11_sufl_tm.csv
   Comments project no.1
   Field data output cycle Constant interval: 0.5 seconds + last cycle
   Initial value for field file Output
   Field file output format Default(FLD file)
   Restart output cycle 100 cycles

Appendix D: Simulation settings for Experiment 4

The following information is the simulations setting for Experiment 1 (take Model J as an example).
* Analysis Types
   Incompressible/Compressible flow Incompressible
   Flow field Turbulent flow
   Turbulence model Standard k-eps model
   Flow Consider
   Free surface MARS method
   Steady Analysis/Transient Analysis Transient analysis
* Basic Settings
   Gravity Consider
   Acceleration due to gravity (0, 0, -1) 9.8 [m/s^2]
* Fluid Region
   Fluid number 1 Domain(cuboid) : air(20C)
   fluid number 2 (VOF2) water(20C)
* Flow
   Components of velocity X, Y, Z-direction
* Initial Condition
   Undefined:Initial temperature of solid parts 20 C
   The number of initial condition settings 1
   water_vol Initial1 : Initial VOF2
* Flow Boundary

　　　　The number of flux boundary settings　2
　　　　Xmax　Flux2 : Natural outflow boundary
　　　　Zmax　Flux3 : Static pressure boundary
* Wall Boundary
　　　　The number of wall boundary settings　3
　　　　Ymin　Wall4 :Freeslip
　　　　Ymax　Wall4 :Freeslip
　　　　Undefined(Stress: All fluid boundary)　Wall2 : Noslip(smooth)
* Symmetrical Boundary
　　　　The number of symmetry boundary settings　None
* Source Condition
　　　　The number of volumetric source condition settings　None
　　　　The number of area source condition settings　None
* Fixed Condition
　　　　The number of fixed velocity condition settings　None
* Free Surface
　　　　Contact angle at wall (entire domain)　90 degrees
　　　　Fractional steps　5 times
　　　　Cut-off value of VOF　0.0001
　　　　Values of VOF under the specified cut-off value　Save
　　　　Control Courant Number in transient analysis　Minimum=0.1 Maximum=0.5
　　　　The number of contact angle condition settings　None
　　　　The number of initial and fixed condition settings　1
　　　　water_vol　Initial1 : Initial VOF2
　　　　The number of permeable body and attenuation zone condition settings　1
　　　　attenuation-zone　FreeSurface5 : Free surface: porous media
　　　　The number of wave generation condition settings　1
　　　　water　wave : Free surface: wave generation
　　　　Output Surface geometry　Output
　　　　Output format　Time series CSV format
　　　　Direction to search free surface　X-direction
* Analysis Control
　　　　Type of setting　Detailed setting
* Transient Analysis
　　　　Cycle　Initial calc.: cycles 1 to 20000
　　　　Time step　Automatically calculated: initial time step=0.0001 Courant no.=0.3

Specified time   360 sec
The number of stop point settings   None
* Solver Parameters
    Matrix solver/Advection term
    X-component of velocity   Default 1st order upwind
    Y-component of velocity   Default 1st order upwind
    Z-component of velocity   Default 1st order upwind
    Pressure   Default ---
    Turbulent kinetic energy   Default 1st order upwind
    Turbulent dissipation rate   Default 1st order upwind
* Field File
    Analysis variable output   Default
* Time Series
    The number of time series settings   None
* List File Output
    Output sum of pressure   1 cycles
    The number of output settings of the sum   10
    Output region   water
    Output region   f1-1
    Output region   f1-2
    Output region   f1-3
    Output region   f2-1
    Output region   f2-2
    Output region   f2-3
    Output region   f3-1
    Output region   f3-2
    Output region   f3-3
    Warning/error message   Output
    Cycle information   Output every cycles
    Matrix calculation information   Output every 1 cycles
* Output Files
    Project name   0_V11
    Generic name for Postprocessor file   0_V11
    Restart file (output)   0_V11.r
    Free surface location file   0_V11_sufl_tm.csv
    Comments   project no.1
    Field data output cycle   Constanct interval: 0.5 seconds + last cycle

Initial value for field file   Output
Field file output format   Default(FLD file)
Restart output cycle   100 cycles

Appendix E: Simulation settings for Experiment 5

The following information is the simulations setting for Experiment 1 (take Model N as an example).

* Analysis Types
    Incompressible/Compressible flow   Incompressible
    Flow field   Turbulent flow
    Turbulence model   Standard k-eps model
    Flow   Consider
    Free surface   MARS method
    Steady Analysis/Transient Analysis   Transient analysis
* Basic Settings
    Gravity   Consider
    Acceleration due to gravity   (0, 0, -1) 9.8 [m/s^2]
* Fluid Region
    Fluid number 1   Domain(cuboid) : air(20C)
    fluid number 2 (VOF2)   water(20C)
* Flow
    Components of velocity   X, Y, Z-direction
* Initial Condition
    Undefined:Initial temperature of solid parts   20 C
    The number of initial condition settings   1
    water_vol   Initial1 : Initial VOF2
* Flow Boundary
    The number of flux boundary settings   2
    Xmax   Flux2 : Natural outflow boundary
    Zmax   Flux3 : Static pressure boundary
* Wall Boundary
    The number of wall boundary settings   3
    Ymin   Wall4 :Freeslip
    Ymax   Wall4 :Freeslip
    Undefined(Stress: All fluid boundary)   Wall2 : Noslip(smooth)
* Symmetrical Boundary

        The number of symmetry boundary settings   None
* Source Condition
        The number of volumetric source condition settings   None
        The number of area source condition settings   None
* Fixed Condition
        The number of fixed velocity condition settings   None
* Free Surface
        Contact angle at wall (entire domain)   90 degrees
        Fractional steps   5 times
        Cut-off value of VOF   0.0001
        Values of VOF under the specified cut-off value   Save
        Control Courant Number in transient analysis   Minimum=0.1 Maximum=0.5
        The number of contact angle condition settings   None
        The number of initial and fixed condition settings   1
        water_vol   Initial1 : Initial VOF2
        The number of permeable body and attenuation zone condition settings   1
        attenuation-zone   FreeSurface5 : Free surface: porous media
        The number of wave generation condition settings   1
        water   wave : Free surface: wave generation
        Output Surface geometry   Output
        Output format   Time series CSV format
        Direction to search free surface   X-direction
* Analysis Control
        Type of setting   Detailed setting
* Transient Analysis
        Cycle   Initial calc.: cycles 1 to 20000
        Time step   Automatically calculated: initial time step=0.0001 Courant no.=0.3
        Specified time   360 sec
        The number of stop point settings   None
* Solver Parameters
        Matrix solver/Advection term
        X-component of velocity   Default 1st order upwind
        Y-component of velocity   Default 1st order upwind
        Z-component of velocity   Default 1st order upwind
        Pressure   Default ---
        Turbulent kinetic energy   Default 1st order upwind
        Turbulent dissipation rate   Default 1st order upwind

* Field File
    Analysis variable output   Default
* Time Series
    The number of time series settings   None
* List File Output
    Output sum of pressure   1 cycles
    The number of output settings of the sum   4
    Output region   water
    Output region   f1
    Output region   f3
    Output region   f2
    Warning/error message   Output
    Cycle information   Output every cycles
    Matrix calculation information   Output every 1 cycles
* Output Files
    Project name   shape
    Generic name for Postprocessor file   shape
    Restart file (output)   shape.r
    Free surface location file   shape_sufl_tm.csv
    Comments   project no.1
    Field data output cycle   Constanct interval: 0.5 seconds + last cycle
    Initial value for field file   Output
    Field file output format   Default(FLD file)
    Restart output cycle   100 cycles

Appendix F: Simulation results of Experiment 1

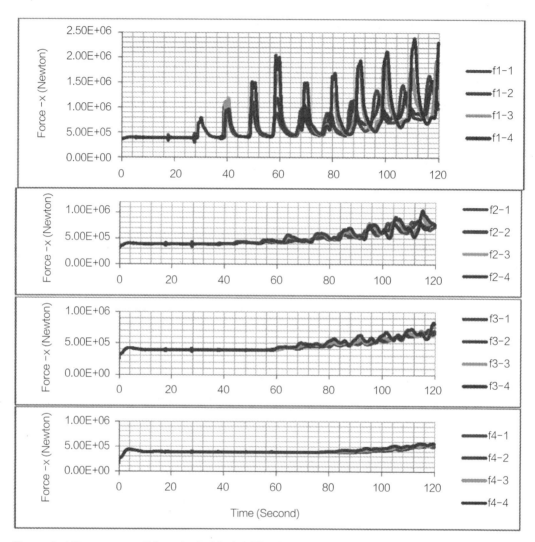

Recorded Force-x on all facadesin Model SP

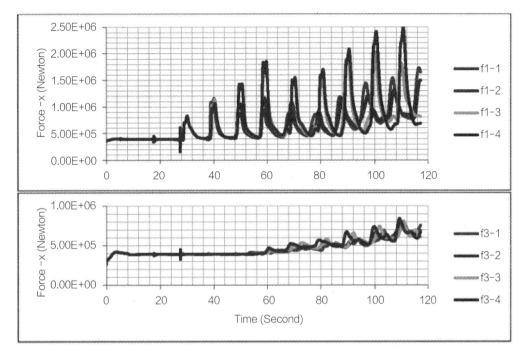

Recorded Force-x on all facades in Model BP

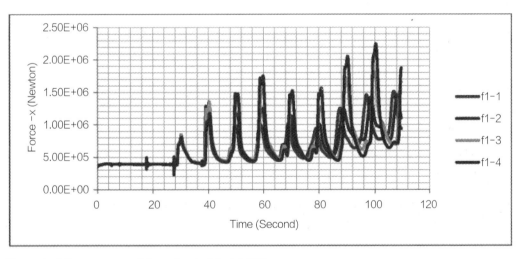

Recorded Force-x on all facades in Model MP

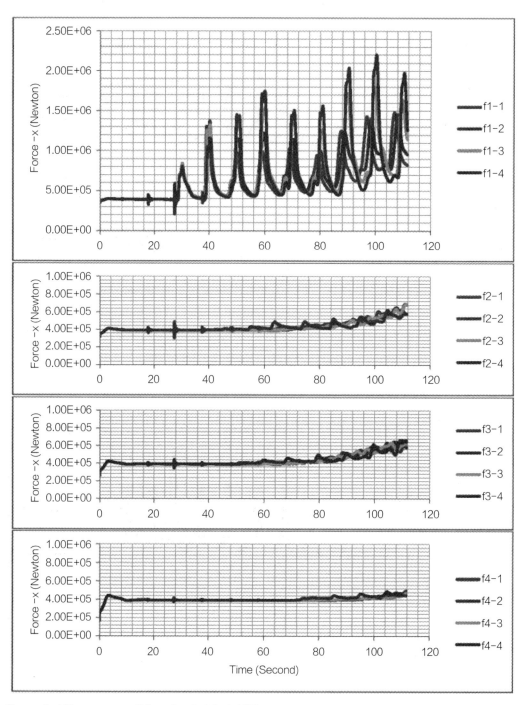

Recorded Force-x on all facades in Model TS

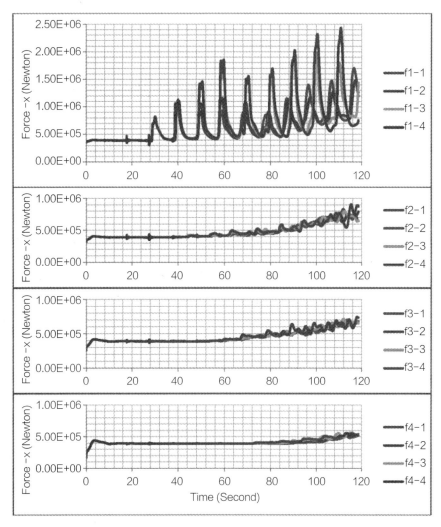

Recorded Force-x on all facades in Model STS

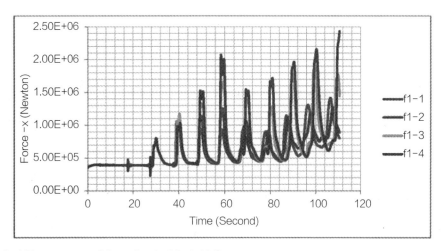

Recorded Force-x on all facades in Model LS

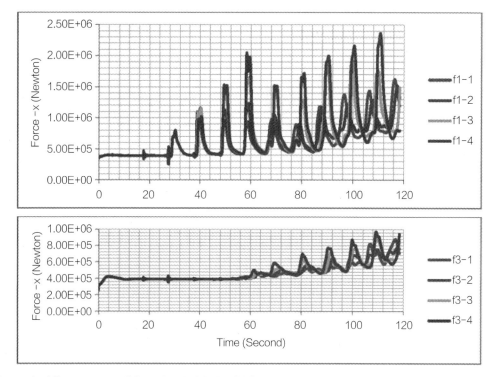

Recorded Force-x on all facades in Model SLS

Appendix G: Simulation results of Experiment 2

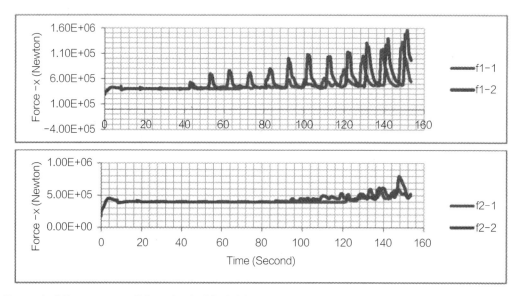

Recorded Force-x on all facades in Model A

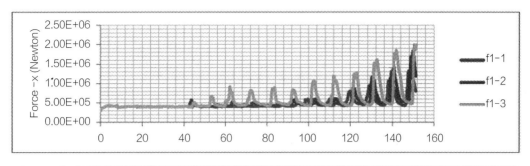

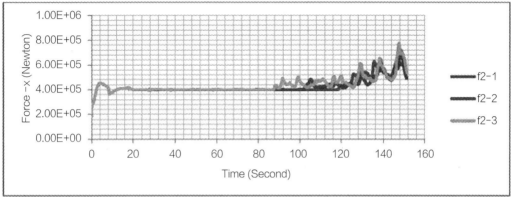

Recorded Force-x on all facades in Model B

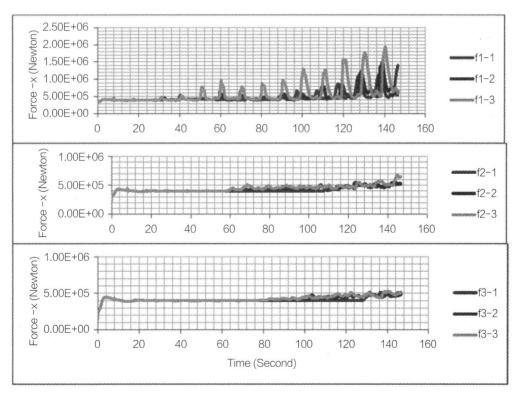

Recorded Force-x on all facades in Model C

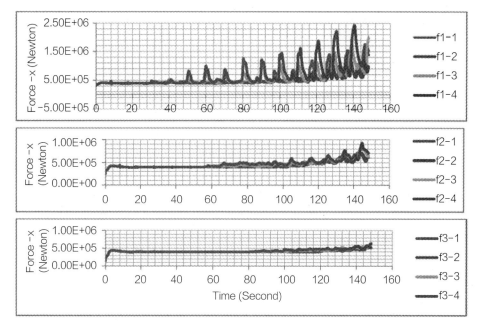

Recorded Force-x on all facades in Model D

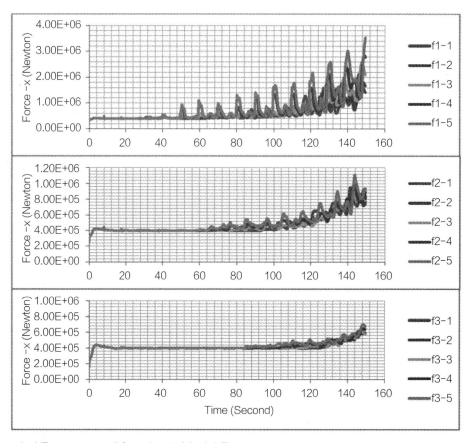

Recorded Force-x on all facades in Model E

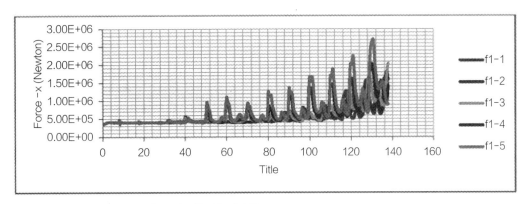

Recorded Force-x on all facades in Model F

Appendix H: Simulation results of Experiment 3

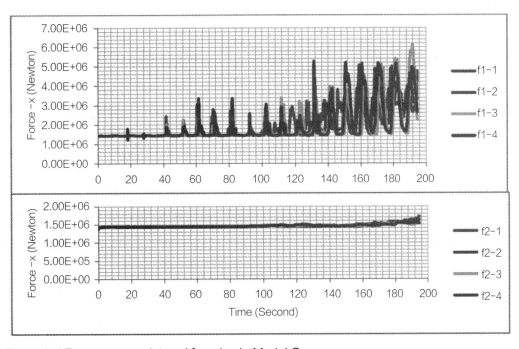

Recorded Force-x on registered facades in Model G

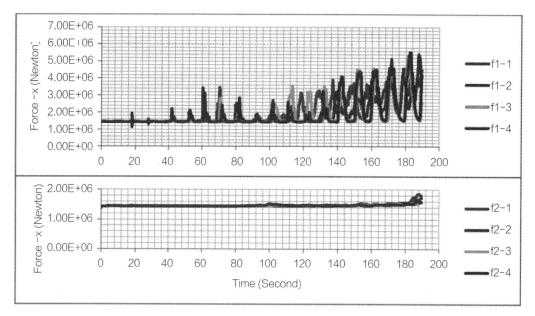

Recorded Force-x on registered facades in Model H

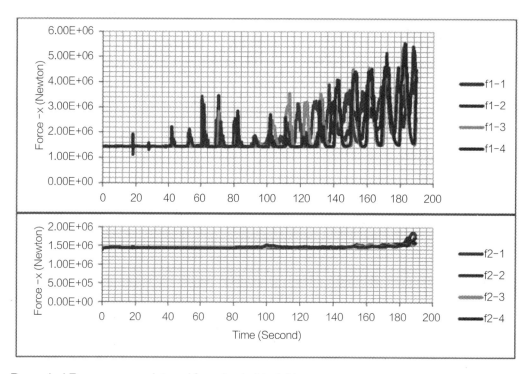

Recorded Force-x on registered facades in Model I

Appendix I: Simulation results of Experiment 4

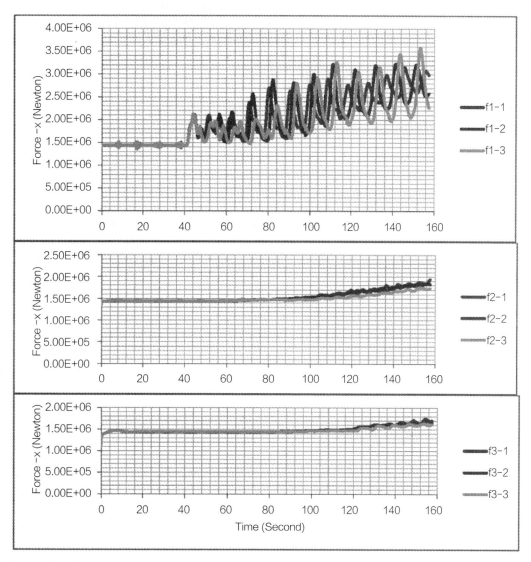

Recorded Force-x on registered facades in Model J

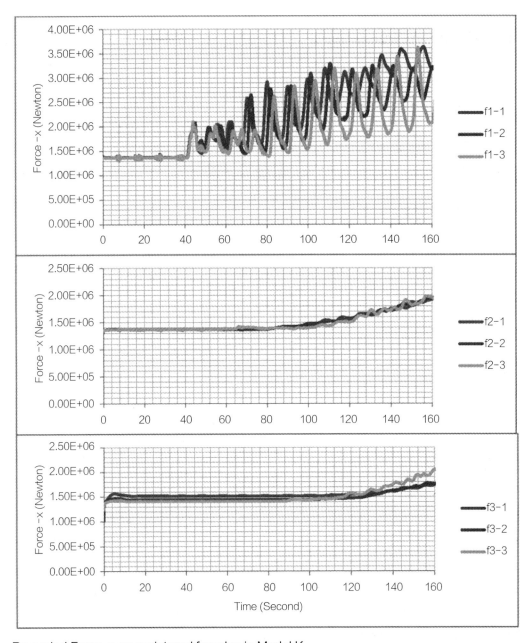

Recorded Force-x on registered facades in Model K

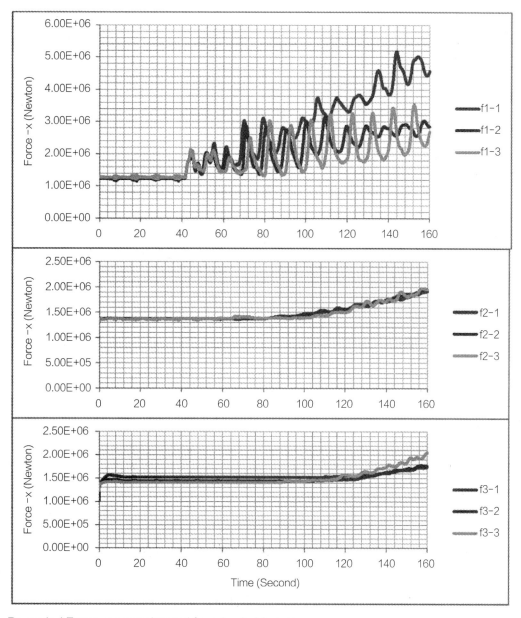

Recorded Force-x on registered facades in Model L

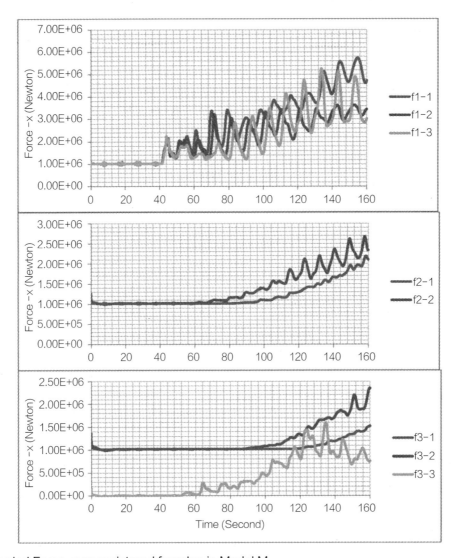

Recorded Force-x on registered facades in Model M

Appendix J: Simulation results of Experiment 5

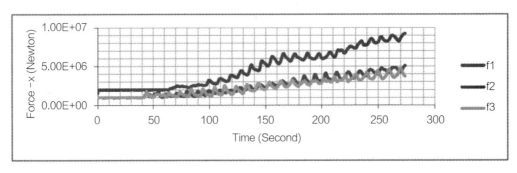

Recorded Force-x on registered facades in Model N

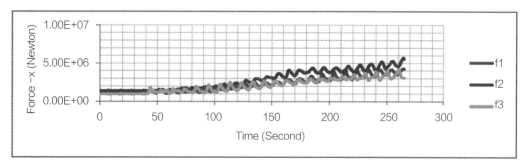

Recorded Force-x on registered facades in Model O

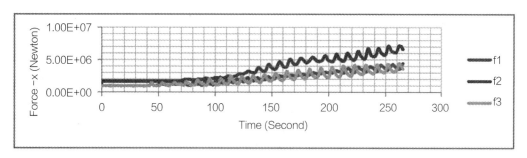

Recorded Force-x on registered facades in Model P

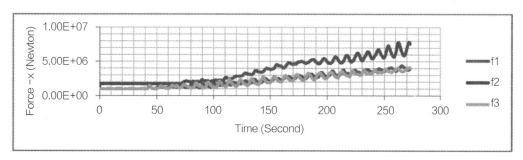

Recorded Force-x on registered facades in Model Q

Appendix K: Simulation settings for Application case study

The following information is the simulations setting for Experiment 1 (take Design 1 as an example).

* Analysis Types

    Incompressible/Compressible flow   Incompressible
    Flow field   Turbulent flow
    Turbulence model   Standard k-eps model
    Flow   Consider
    Free surface   MARS method

Steady Analysis/Transient Analysis   Transient analysis
* Basic Settings
    Gravity   Consider
    Acceleration due to gravity   (0, 0, -1) 9.8 [m/s^2]
* Fluid Region
    Fluid number 1   Domain(cuboid) : air(20C)
    fluid number 2 (VOF2)   water(20C)
* Flow
    Components of velocity   X, Y, Z-direction
* Initial Condition
    Undefined:Initial temperature of solid parts   20 C
    The number of initial condition settings   1
    water_vol   Initial1 : Initial VOF2
* Flow Boundary
    The number of flux boundary settings   2
    Xmax   Flux2 : Natural outflow boundary
    Zmax   Flux3 : Static pressure boundary
* Wall Boundary
    The number of wall boundary settings   3
    Ymin   Wall4 :Freeslip
    Ymax   Wall4 :Freeslip
    Undefined(Stress: All fluid boundary)   Wall2 : Noslip(smooth)
* Symmetrical Boundary
    The number of symmetry boundary settings   None
* Source Condition
    The number of volumetric source condition settings   None
    The number of area source condition settings   None
* Fixed Condition
    The number of fixed velocity condition settings   None
* Free Surface
    Contact angle at wall (entire domain)   90 degrees
    Fractional steps   5 times
    Cut-off value of VOF   0.0001
    Values of VOF under the specified cut-off value   Save
    Control Courant Number in transient analysis   Minimum=0.1 Maximum=0.5
    The number of contact angle condition settings   None
    The number of initial and fixed condition settings   1

water_vol   Initial1 : Initial VOF2

The number of permeable body and attenuation zone condition settings   1

attenuation-zone   FreeSurface5 : Free surface: porous media

The number of wave generation condition settings   1

water   wave : Free surface: wave generation

Output Surface geometry   Output

Output format   Time series CSV format

Direction to search free surface   Z-direction

The number of settings of the locations to begin search for free surface   None

* Analysis Control

Type of setting   Detailed setting

* Transient Analysis

Cycle   Initial calc.: cycles 1 to 15000

Time step   Automatically calculated: initial time step=0.001 Courant no.=0.3

Specified time   300 sec

The number of stop point settings   None

* Solver Parameters

Matrix solver/Advection term

X-component of velocity   Default 1st order upwind

Y-component of velocity   Default 1st order upwind

Z-component of velocity   Default 1st order upwind

Pressure   Default ---

Turbulent kinetic energy   Default 1st order upwind

Turbulent dissipation rate   Default 1st order upwind

* Field File

Analysis variable output   Default

* Time Series

The number of time series settings   None

* List File Output

Output sum of pressure   1 cycles

The number of output settings of the sum   46

Output region   water

Output region   1-1

Output region   1-2

Output region   1-3

Output region   1-4

Output region 2-1
Output region 2-2
Output region 2-3
Output region 3-1
Output region 3-2
Output region 4-2
Output region 2-4
Output region 4-3
Output region 5-1
Output region 5-2
Output region 6-1
Output region 6-2
Output region 6-3
Output region 6-4
Output region 7-1
Output region 7-2
Output region 7-3
Output region 7-4
Output region 11-1
Output region 11-2
Output region 11-3
Output region 11-4
Output region 10-1
Output region 10-2
Output region 10-3
Output region 10-4
Output region 10-5
Output region 9-3
Output region 9-1
Output region 9-2
Output region 8-1
Output region 8-2
Output region 8-3
Output region 8-4
Output region 8-6
Output region 8-7
Output region 8-8

　　　　　Output region　8-5
　　　　　Output region　3-3
　　　　　Output region　4-1
　　　　　Output region　12-1
　　　　　Warning/error message　Output
　　　　　Cycle information　Output every cycles
　　　　　Matrix calculation information　Output every 1 cycles
* Output Files
　　　　　Project name　before3
　　　　　Generic name for Postprocessor file　before3
　　　　　Restart file (output)　before3.r
　　　　　Comments　　1000-800-100, wd 8 wh 20 wp 5
　　　　　Field data output cycle　Constanct interval: 0.5 seconds + last cycle
　　　　　Initial value for field file　Output
　　　　　Field file output format　Default(FLD file)
　　　　　Restart output cycle　100 cycles

Appendix L: Simulation results of Design 1

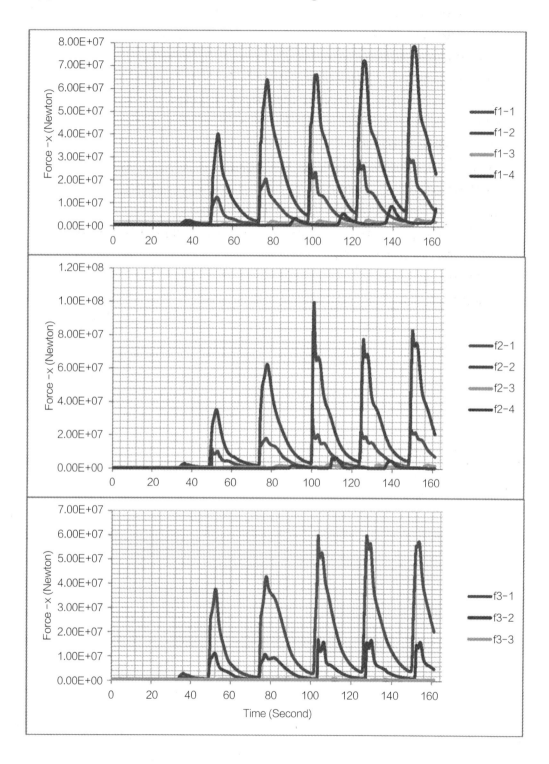

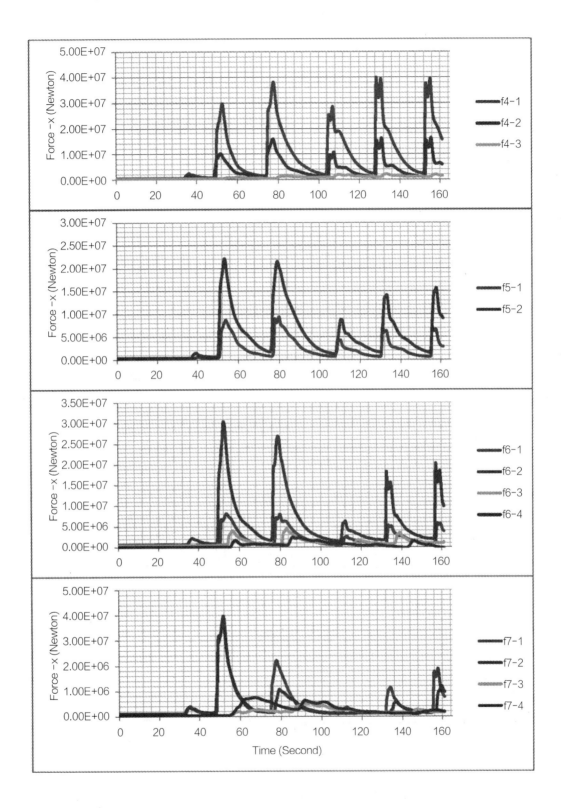

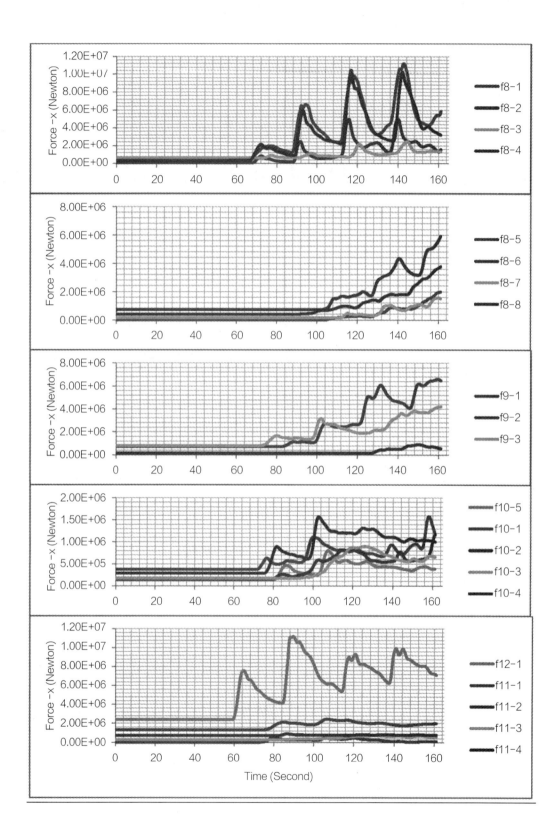

Appendix M: Simulation results of Design 2

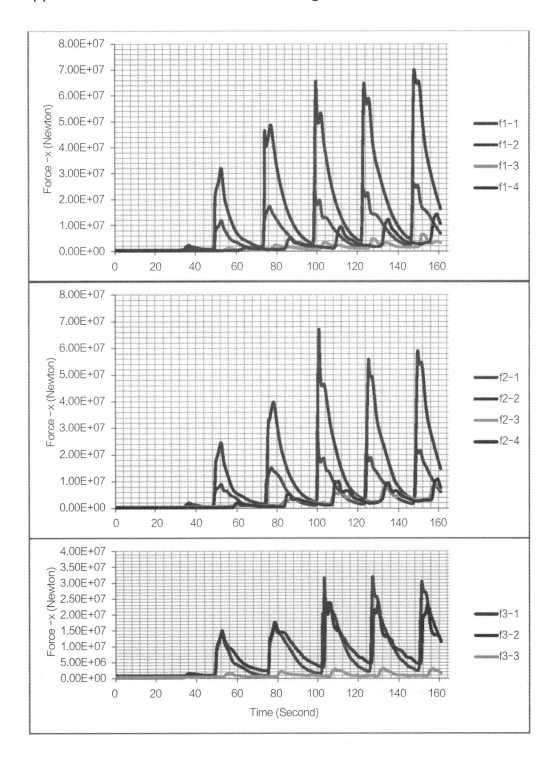

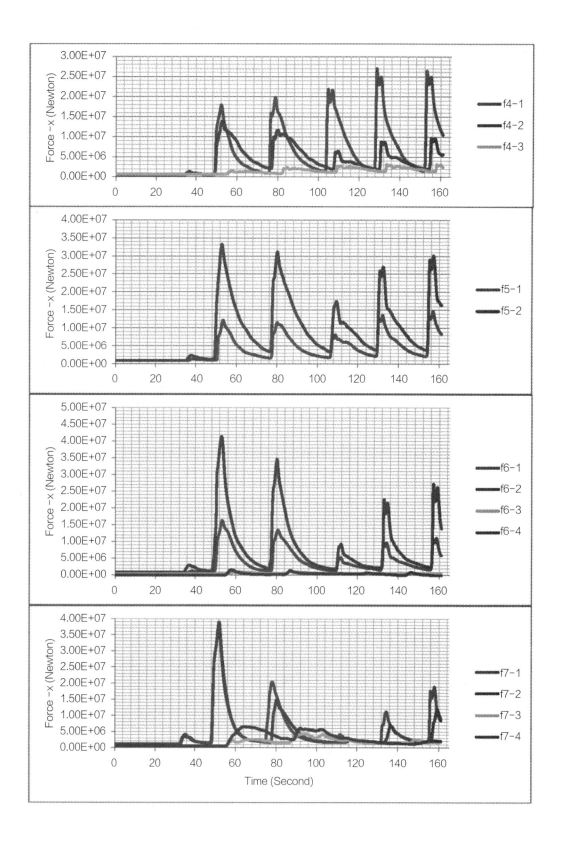

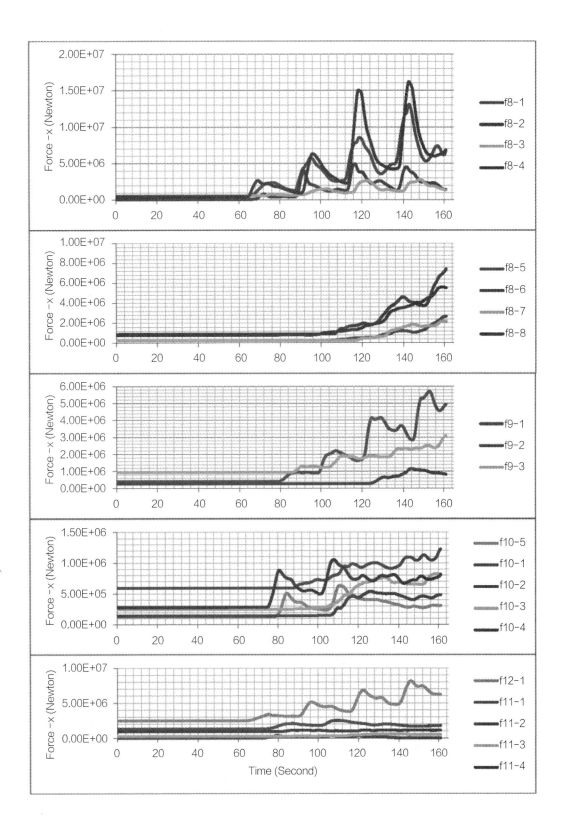